SUPERCONDUCTIVITY RESEARCH AND APPLICATIONS

SUPERCONDUCTIVITY AND CONDENSED MATTER RESEARCHERS

BIOGRAPHICAL SKETCHES AND RESEARCH SUMMARIES

SUPERCONDUCTIVITY RESEARCH AND APPLICATIONS

Additional books in this series can be found on Nova's website under the series tab.

Additional e-books in this series can be found on Nova's website under the e-book tab.

SUPERCONDUCTIVITY RESEARCH AND APPLICATIONS

SUPERCONDUCTIVITY AND CONDENSED MATTER RESEARCHERS

BIOGRAPHICAL SKETCHES AND RESEARCH SUMMARIES

COLLIN J. ECKLES
EDITOR

Nova Science Publishers, Inc.
New York

For permission to use material from this book please contact us:
Telephone 631-231-7269; Fax 631-231-8175
Web Site: http://www.novapublishers.com

Library of Congress Cataloging-in-Publication Data

Superconductivity and condensed matter researchers: biographical sketches and research summaries / editor, Collin J. Eckles.
p. cm.
ISBN 978-1-61209-785-5 (hardcover)
1. Superconductors. 2. Condensed matter. 3. Superconductivity--Research.
4. Materials scientists--Biography. I. Eckles, Collin J.
QC611.95.S88 2011
537.6'23--dc22 2011003588

Published by Nova Science Publishers, Inc. † New York

CONTENTS

PREFACE

This new book compiles biographical sketches of top professionals in the fields of superconductivity and condensed matter, as well as research summaries from a number of different focuses in this important field.

PART 1 - RESEARCH BIOGRAPHIES

In: Superconductivity and Condensed Matter … ISBN 978-1-61209-785-5
Editor: Collin J. Eckles

Chapter 1

BIOGRAPHICAL SKETCH – Superconductivity Research

NAME: EL-METWALLY MAHMOUD ABEDEL RAZEK	TITLE: Physics Department, Faculty of Science, Mansoura University, Mansoura, P.O. BOX 35516, EGYPT	DATE OF BIRTH: 19-11-1962

EDUCATION

Institution and Location	Degree	Year Conferred	Scientific Field
Faculty of Science, Mansoura University, Mansoura, Egypt	B. Sc. Degree	May 1985	Physics by grade "Very good".
Faculty of Science, Mansoura University, Mansoura, Egypt	M. Sc. Degree	August 27, 1990	Physics (Solid state Physics)
Faculty of Science, Mansoura University, Mansoura, Egypt	Ph. D. of Science	February 27, 1995	Physics (Solid state Physics)

RESEARCH AND PROFESSIONAL EXPERIENCE

* Effect of fillers on Physical properties of polymers
* Effect of Inorganic Materials different concentrations on the physical properties and structure of Some Polymers

Publications during Last Three Years

[1] E.M. Abdelrazek, I.S. Elashmawi, Characterization and Physical Properties of $CoCl_2$ Filled Polyethyl-methacrylate Films, Polymer Composites 29 (2008) 1036.

[2] E.M. Abdelrazek, Spectroscopic studies on the effect of doping with $CoBr_2$ and $MgCl_2$ on some physical properties of polyvinylalcohol films, Physica B 403 (2008) 2137.

[3] I. S. Elahmawi, N. A. Hakeem, E. M. Abdelrazek, Miscibility and spectroscopic studies of PS/PVAc blends. Physica B 403 (2008) 3547.

[4] E.M. Abdelrazek, I.S. Elashmawi, H.M. Ragab, Manifestation of MnCl2 fillers incorporated into the polymer matrices in their dielectric properties, Physica B 403 (2008) 3097.

[5] E. M. Abdelrazek, I. S. Elashmawi, M. A. Soliman and A. Aly, Physical properties of MnCl2 fillers incorporated into PVDF/PVC blend and their complexes, Journal of Vinyl and Additive Technology, 15 (2009) 171

[6] I. K. El Zawawi, E. M. Abdelrazek, F. A. Saadallah, B. Mansour, M. Ebrahim, F. Muecklich, C. Holzapfel, H.A. Gobran, Study of phase evolution in sputtered Al/Ru bi-layers nanocrystalline thin films, Egypt. Journal Solids, 32 (2009) 89.

[7] E. M. Abdelrazek, I.S. Elashmawi, A. El-khodary, A. Yassein, Structural, optical, thermal and electrical studies on PVA/PVP blends filled with lithium bromide, Current Applied Physics, 10 (2010) 607

[8] I.S. Elashmawi, E.M. Abdelrazek, Role of $CoBr_2$ on the structural, optical and magnetic properties of polyvinyl alcohol films. Journal of Applied Polymer Science, 115 (2010) 2691.

[9] I.S. Elashmawi, E.M. Abdelrazek, H.M. Ragab, N.A. Hakeem, Structural, optical and dielectric behavior of PVDF films filled with different concentrations of iodine, Physica B 405 (2010) 94.

[10] E. M. Abdelrazek, G. M. El Damrawi, A. Al-Shahawy, Some studies on Calcium Phosphate Embedded in PVA Matrix before and after γ-irradiation, PhysicaB 405(2010) 808.

[11] E. M. Abdelrazek, G. El Damrawi, I. S. Elashmawi, A. El-Shahawy, The influence of γ-irradiation on Some Physical properties of Chlorophyll/PMMA films, Applied Surface Science 256 (2010) 2711.

[12] R. Seoudi, M. Kamal, A.A. Shabaka, E.M. Abdelrazek, W. Eisa, Synthesis, characterization and spectroscopic studies of CdS/polyaniline core/shell nanocomposite, Synthetic Metals 160 (2010) 479.

[13] E.M. Abdelrazek, I.S. Elashmawi, S. Labeeb, Chitosan filler effects on the experimental characterization, spectroscopic investigation and thermal studies of PVA/PVP blend films, Physica B 405 (2010) 2021

[14] E.M. Abdelrazek, Hosam S. Ibrahim, Effect of heparin calcium different concentrations on some physical properties and structure in polyacrylamide matrix , Physica B (In press, online 24-7-2010)

Other contributions of interest to "Effect of Inorganic Materials different concentrations on the physical properties and structure of Some Polymers" research: Projects, Grants etc.

In: Superconductivity and Condensed Matter ... ISBN 978-1-61209-785-5
Editor: Collin J. Eckles

Chapter 2

BIOGRAPHICAL SKETCH – Superconductivity Research

NAME: HIND IBRAHEEM ABDULGAFOUR AL SHAIKH	TITLE: Researcher	DATE OF BIRTH: 18-2-1967 – BAGHDAD, IRAQ
Marital Status: Married	Nationality: Iraqi	

CONTACT POINTS

E-Mail: hind_alshaikh@ yahoo.com
Postal address: School of Physics, Universiti Sains Malaysia (USM), 11800 Penang, Malaysia
Tel.: +60174500286

PRESENT OCCUPATION: CHIEF OF PHYSICISTS (HEAD OF SECTION)

EDUCATION AND TRAINING

QUALIFICATIONS

- PhD student, School of Physics, Universiti Sains Malaysia (USM)
- MSc., Physics, Baghdad University, 1999.
- BSc., Physics, Baghdad University, 1988.
- A course in English language at American Global English Institute at July, 7, 2005.
- A traning of Comprehensive National Hazardous Material Management Program held in America 2 August -16 August 2005.
- A course in fine membrane properties.
- A course in environment protection from industrial pollution, held by the The Arabic Union for Engineering 21 Dec. – 26 Dec. 2002.

SCIENTIFIC EXPERIENCE

1. Worked at Atomic Energy Commission in 1988-2003.
2. Head Section of Air Pollution Technology at Atomic Energy Commission in 2001.
3. Worked at Ministry of Scince And Technology from 2003 until now.
4. Currently, PhD student in Universiti Sains Malaysia (USM). The title of my study is A Study on the Fabrication and Characterization of synthesis Zinc Oxide Nanostructures.
5. Published (15) scientific researches, (7) as main researcher, (8) as participant.
6. Design and construction of laser system to detect air pollutants for factories, and determination of pollutants size and concentration.
7. Development and manufacturing of fabric filters (catridge type) for filtration of industrial water used in various industrial sectors such as a electricity and oil sectors.
8. Got a patent entitled Preparation of Wax Polymer Mixtures and A Study Of Physical Properties for These Materials.
9. Design and manufacturing of industrial pollution removing system from the air in closed area.
10. Preparation of polystyrene used in medical appliance.
11. Participating in scientific conferences held in Iraq, especially those prepared by physics & mathematics society, and delivering many research papers in these conferences

SCIENTIFIC PUBLICATIONS

Published (15) scientific researches, (7) as main researcher, (8) as participant , including:

1 H.I. Abdulgafour,Z. Hassan, N.H. Al-Hardan, F.K. Yam, " Growth Of High-Quality Nanowires Without Catalyst" , j.physb.2010.07.013

2 H.I. Abdulgafour,Z. Hassan, N.H. Al-Hardan, F.K. Yam, "Growth Of Zinc Oxide Nanoflowers By Thermal Evaporation Method. j.physb. Physica B 405 (2010) 2570–25722.

3 Hind.I.Abdulgafour,Hussain.K.Sulaiman, Rehabilitation Of Tissues Used In Artificial Kidney Washing ,J.College Of Education / Al_Mustansiriyah University ,No.3, 82_90,2002.

4 Hind.I.Abdulgafour,Ehsaan.Y.Naseem,Tamer.A.Hadi, Design And Manufacturing Of Industrial Pollution & Radiation Particles Removing System From The Air In Closed Area, J.College Of Education / Al_Mustansiriyah University ,No.1, 319_324,2002.

5 Hind.I.Abdulgafour,Tamer.A.Hadi, Real & Imaginary Isolation Factor Measurement For Cement Mixtures, J.College Of Education / Al_Mustansiriyah University ,No.5,295_202, 2001.

6 Hind.I.Abdulgafour,Rafe.K.Al_Kubaisy,Tamer.A.Hadi , A Study For The Electrical Properties For Paraffin Mixtures With High Dencity Polly-Ethylene (HOPE)& Low Density (LEPE) For Frequency Range Of (10-1000) KHz,& The Effect Of Gamma Ray On These Properties, , J.College Of Education / Al_Mustansiriyah University ,No.5,225_238,2001.

7 Hind.I.Abdulgafour,Tamer.A.Hadi, Possibility Of Properties For Various Type Improving Physical Es Of Iraqi Oil, , J.College Of Education / Al_Mustansiriyah University, No.9,343_346,2001.

8 Rafe.K.Al_Kubaisy,Hind.I.Abdulgafour, A Study For Electrical & Mechanical Properties For Polymer Wax Arrys & The Effect Of Gamma Ray On These Properties , Sci.J.Iraqi Energy Commission /Vol.2,No.2 ,2000.
9 Tamer.A.Hadi,Hind.I.Abdulgafour,Ehsaan.Y.Naseem Evaluation Of City Aia Type, J.College Of Education / Al_Mustansiriyah University ,No. 9,387_398,2001.
10 Faleh.H.Hamza ,Ahmad.S.Wasfi ,Hind.I.Abdulgafour,Design And Costruction Of (TEA CO2) Laser System, J.College Of Education / Al_Mustansiriyah University ,No.6, 6_10, 2001.
11 Tamer.A.Hadi,Hind.I.Abdulgafour,Ehsaan.Y.Naseem. Development Of Water Filter For Industrial Water (Cartrdige Type) To Be Used In Various Industrial Sectors, , J.College Of Education / Al_Mustansiriyah University ,No.6, 6_10, 2001.
12 Tamer.A.Hadi,Hind.I.Abdulgafour,Determination Of Alternative Raw Materails For Air Filters From Industrial Pollutants, J.College Of Education / Al_Mustansiriyah University, No.3,49-56 , 2001.
13 Tamer.A.Hadi,Hind.I.Abdulgafour, Determination Of Physical Properties Of Industrial Water Filters Made Of Cotton Threads Using Raw Materials And Various Engineering Shapes . J.College Of Education / Al_Mustansiriyah University ,No.1, 361-364,2002.
14 Tamer.A.Hadi,Hind.I.Abdulgafour,Production Of Fabric Filters To Clean The Air From Industrial Pollutants And Dust, J.College Of Education / Al_Mustansiriyah University ,No.1, 299-302,2002.
15 Fuad.H.Kamel,Rafe.K.Al_Kubaisy,Hind.I.Abdulgafour Preparation Of Polystyrene Used In Medical Appliance J.College Of Education / Al_Mustansiriyah University ,No.3, 40-48,2002.

MAIN STUDIES

1. A Study on the Fabrication and Characterization of synthesis Zinc Oxide Nanostructures for sensor application.
2. Construction and manufacturing of laser system for the measurement of air pollutants.
3. Determination of filters specifications suitable for various industrial activities.
4. Determination of size and concetration of air pollutants using diffusion by laser ray.

In: Superconductivity and Condensed Matter ... ISBN 978-1-61209-785-5
Editor: Collin J. Eckles

Chapter 3

BIOGRAPHICAL SKETCH – Superconductivity Research

NAME:	TITLE:	DATE OF BIRTH:
SASHA ALEXANDROV	PROFESSOR	30/07/1946

EDUCATION

Institution and Location	Degree	Year Conferred	Scientific Field
Moscow Engineering Physics Institute	M.S., Theoretical Physics (1st class honor)	1970	Condensed Matter Physics
Moscow Engineering Physics Institute	Ph.D., Theoretical Physics	1973	Condensed Matter Physics
Moscow Engineering Physics Institute	Doctor of Sciences., Theoretical	1984	Condensed Matter Physics

RESEARCH AND PROFESSIONAL EXPERIENCE

PRESENT POSITION

Professor of Theoretical Physics, Loughborough University, United Kingdom

PREVIOUS POSITIONS

1973-1990 Lecturer, associate professor, full professor, Dean of Theoretical and Experimental Physics Faculty, Head of General Physics Department, Director of Superconductivity Research Laboratory, Vice Rector of *Moscow Engineering Physics Institute*, Moscow, Russia

1990-1992 Visiting Professor, *Rhein –Westfalische Technische Hochschule* (RWTH) Aachen, Germany
1992-1995 Research Professor, *IRC in Superconductivity, University of Cambridge*, UK
1995-1998 Chair in Theoretical Physics, *Department of Physics, Loughborough University*, UK
1998-2001 Head of Physics Department, *Loughborough University*, UK
2001-to date Professor of Theoretical Physics, *Loughborough University*, UK

PROFESSIONAL APPOINTMENTS

Since 1996 Visiting Professor, *IRC in Superconductivity, University of Cambridge*, United Kingdom
Since 1998 External Consultant and Visiting Professor at *Hewlett Packard Research Laboratories*, Palo Alto, California, United States
2008 Visiting Professor at Jozef Stefan Institute, Ljubljana, Slovenia

HONORS RELEVANT TO SUPERCONDUCTIVITY RESEARCH

1983-1989 Moscow Engineering Physics Institute Prizes, Ministry of Atomic Energy Prizes (USSR) for achievements in research
1992 Mott Bye-Fellow (Gonville and Caius College, Cambridge, UK) (first awarded)
1996 Fellow of IoP (London)
2003 NATO-country co-director of the Advanced Research Workshop "Molecular Nanowires"
2004 Guest Editor of a special volume of Physica C Superconductivity (The Netherlands)
2005 Scientific Convenor of the European Science Foundation Workshop "Mott's Physics in Nanowires and Quantum Dots"
2007 Guest Editor of a special volume of Journal of Physics: Condens. Matter
2008 Slovenian Research Agency (ARRS) award for outstanding foreign researchers
2008 Fellow of the American Physical Society (elected for "important and broad-ranging contributions to the theory of correlated electron systems").
2009 Lead Guest Editor of a special issue of Advances in Condensed Matter Physics

NAMED LECTURES

1997 "Enrico Fermi" lecture (International School of Physics, Course CXXXVI, Varenna, Italy)
2000 "Mott Lecture" (Florida State University and MARTECH, (Tallahassee, Florida, US) (1st Annual lecture)

Invited Speaker at 75 international conferences, workshops and colloquia

PUBLICATIONS DURING LAST THREE YEARS

1. *Alexandrov AS, Beanland, J, *Superconducting Gap, Normal State Pseudogap, and Tunneling Spectra of Bosonic and Cuprate Superconductors.* PHYS. REV. LETT. 104 (2): 026401 JAN 2010
2. Alexandrov AS, Bratkovsky AM, *Giant thermopower and figure of merit of semiconducting polaronic nanolayers,* Phys. Rev. B 81: 153204 APR 2010
3. Alexandrov AS, Bratkovsky AM, *Polaronic memory resistors strongly coupled to electrodes*, Phys. Rev. B 80: 115321 SEP 2009
4. Demsar, J, Kabanov, VV, Alexandrov, AS, *et al., Hot electron relaxation in the heavy-fermion Yb1-xLuxAl3 compound using femtosecond optical pump-probe spectroscopy,* PHYS. REV. B 80: 085121 AUG 2009
5. Hardy TM, Hague JP, Samson JH, Alexandrov AS, *Superconductivity in a Hubbard-Froumlhlich model and in cuprates*, PHYS. REV. B 79: 212501 JUN 2009
6. Alexandrov AS, Devreese J.T., *Advances in Polaron Physics,* Springer Series in Solid-state Sciences (2009)
7. Alexandrov AS , *Phase Separation of Electrons Strongly Coupled with Phonons in Cuprates and Manganites*, Journal Of Superconductivity And Novel Magnetism 22: 95 (2009)
8. Alexandrov AS, Tomas IO, *Zero-temperature phase transitions in dilute bosonic superfluids on a lattice,* J. Phys.: Condens. Matter 21: 245602 (2009)
9. Devreese J.T And Alexandrov AS, *Fröhlich polaron and bipolaron: recent developments,* REP. PROG. PHYS. 72: 066501 (2009)
10. Alexandrov AS, *Unconventional Pairs Glued by Conventional Phonons in Cuprate Superconductors,* Journal Of Superconductivity And Novel Magnetism 22: 103 (2009)
11. Alexandrov AS, *Bipolaronic proximity and other unconventional effects in cuprate superconductors* in Electron Transport In Nanosystems (NATO Advanced Research Workshop) p. 139 (2008)
12. Alexandrov AS, *Angle-resolved photoemission spectroscopy of band tails and anomalous kinetics of underdoped cuprates*, Journal Of Physics And Chemistry Of Solids 69 : 2971 (2008)
13. Hague JP, Kornilovitch PE, Samson J, Alexandrov AS, *Singlet and triplet bipolarons on the triangular lattice*, Journal Of Physics And Chemistry Of Solids 69: 3304 (2008)
14. Kabanov VV, Alexandrov AS, *Electron Relaxation in Metals: Theory and Exact Analytical Solutions*, PHYS. REV B 78: 174514 (2008)
15. Alexandrov AS, Kabanov VV, *Low-temperature proximity effect in clean metals with repulsive electron-electron interaction*, Phys. Rev. B 78: 132510 (2008)
16. Hague JP, Kornilovitch PE, Alexandrov AS, *Trapping of lattice polarons by impurities,* Phys. Rev B 78: 092303 SEP 2008
17. Alexandrov AS, *Theory of quantum magneto-oscillations in underdoped cuprate superconductors,* J Phys-Condens Mat 20 (19) 192202 May 2008
18. Kabanov VV, Alexandrov AS, *Magnetic quantum oscillations in doped antiferromagnetic insulators,* Phys. Rev B 77: 1325403 Apr 2008
19. *Alexandrov AS, *Unconventional pairing symmetry of layered superconductors caused by acoustic phonons,* Phys. Rev B 77: 094502 Mar 2008
20. Thomas IO, Kabanov VV, Alexandrov AS, *Shubnikov-de Haas effect in multiband quasi-two-dimensional metals,* Phys. Rev B 77: 075434 Feb 2008

21. Alexandrov AS, Kabanov VV, *Combination quantum oscillations in canonical single-band Fermi liquids,* Phys. Rev B 76: 233101 Dec 2007
22. Alexandrov AS, Kabanov VV, Thomas IO, *Interplay of size and Landau quantizations in the de Haas-van Alphen oscillations of metallic nanowires,* Phys. Rev B 76: 155417 Oct 2007
23. Alexandrov AS, *Bose-Einstein Condensation in the Pseudogap Phase of Cuprate Superconductors,* J. Supercond Nov Magn 20 (7-8) 481 Nov 2007
24. Alexandrov AS, *Superconducting polarons and bipolarons,* In *Polarons in Advanced Materials* (ed. A.S. Alexandrov, Spriner) SEP 2007, PP 257-310, ISBN 978-1-4020-6347-3
25. Alexandrov AS, *Strong-coupling theory of high-temperature superconductivity,* In *Studies of High Temperature Superconductivity, Golden Jubilee Volume* (ed. A. Narlikar, Nova Science) Jan 2007, Pp 1-70, ISBN 1-59454-960-5
26. Alexandrov AS, *D-wave checkerboard Bose condensate of mobile bipolarons,* Int. J. Mod. Phys. B 21 (13/14) 2301 May 2007
27. Alexandrov AS, *Theory of giant and nil proximity effects in cuprate semiconductors,* Phys. Rev B 75: 132501 Apr 2007
28. Alexandrov AS, Bratkovsky AM, *Fast polaron switching in degenerate molecular quantum dots,* J Phys-Condens Mat 19 (25) 255203 Jun 2007
29. Hague JP, Kornilovitch PE, Samson JH and Alexandrov AS, *Superlight small bipolarons,* J Phys-Condens Mat 19 (25) 255214 Jun 2007
30. Alexandrov AS , *Bose-Einstein condensation of strongly correlated electrons and phonons in cuprate superconductors,* J Phys-Condens Mat 19 (12) 125216 Mar 2007
31. Alexandrov AS, *Superlight Small Bipolarons: A Route to Room Temperature Superconductivity,* In *High Tc Superconductors and Related Transition Metal Oxides* (eds. A. Bussmann-Holder and H. Keller; Springer-Verlag Berlin Heidelberg) Apr 2007, PP 1-15, ISBN 978-3-540-71023
32. Hague JP, Kornilovitch PE, Alexandrov AS and Samson JH., *Unconventional pairing in bipolaronic theories,* Physica C: 460, 1121 SEP 2007
33. *Hague JP, Kornilovitch PE PE, Samson JH. and Alexandrov AS, *Superlight Small Bipolarons in the Presence of a Strong Coulomb Repulsion* Phys. Rev. Lett. 98(3): 037002 JAN 2007

OTHER CONTRIBUTIONS OF INTEREST TO SUPERCONDUCTIVITY RESEARCH: PROJECTS, GRANTS, ETC.

GRANTS HELD

- The Royal Society grant for collaboration with Jozef Stefan Institute (Slovenia) on ultrafast pump-probe spectroscopy of complex matter (2009-2011)
- The Leverhulme Trust visiting professorship award for collaboration with Kapitza Institute for Physical Problems (2010)
- Collaboration with Brookhaven National Laboratory on Novel Superconductors (EPSRC)
- Magnetic Quantum Oscillations in Nanowires (EPSRC), PI 2006-2009
- Polaron Dynamics and Isotope Effects In Complex Oxides (EPSRC), PI 2005-2007

- Mott's Physics in Nanowires and Quantum Dots, (ESF Exploratory Workshop EW05-106)
- Mott's Physics in Nanowires and Quantum Dots, (EPSRC support for above workshop)

In: Superconductivity and Condensed Matter …
Editor: Collin J. Eckles
ISBN 978-1-61209-785-5

Chapter 4

BIOGRAPHICAL SKETCH – Superconductivity Research

NAME:	TITLE:	DATE OF BIRTH:
ALEJANDRO CABO MONTES DE OCA	PHD	19 SEPTEMBER 1947

EDUCATION

Institution and Location	Degree	Year Conferred	Scientific Field
Faculty of Physics, Havana University, Havana	Undergraduate studies of Physics	1970	Physics
Faculty of Physics, Havana University, Havana	MSc Dissertation	1976	Paramagnetic Resonance Physics
P. N. Lebedev, Physical Institute, Moscow	PhD Thesis	1982	Gauge Field Theory, Theoretical Physics

CONTACT POINTS

Alejandro Cabo Montes de oca
Departamento de Fisica Teorica
Instituto de Cibernetica, Matematica y Fisica
Calle E, No. 309, entre 13 y 15, Vedado, La Habana, Cuba
Tel. 537 8303370
Emails: cabo@icmf.inf.cu y alejacabo@gmail.com

Research and Professional Experience

Membership in Scientific Organizations

1. Member of the Cuban Physical Society (1970-2009)
2. Member of the American Physical Society (Actual member, and for various years after 1980)
3. Senior Associate Member of the International Centre for Theoretical Physics (IAEA and UNESCO) for the 1997-2003 period.
4. Regular Associate Member of the International Centre for Theoretical Physics (IAEA and UNESCO) for the 1991-1996 period.
5. Member of the Higher Scientific Council of the Cuban Academy of Sciences for the 1989 - 1992 period.
6. Member of the Scientific Council of the Instituto de Cibernética, Matemática y Física since 1983-2009.
7. Associate Member CEFIMAS 2005-2006 (Buenos Aires, Argentina).

Formation of Other Specialists

A number of nearly 23 Graduate, Master and PhD dissertation works had been supervised.

Special Recognitions

1. *Carlos J. Finlay Medal*, Consejo de Estado, Republica de Cuba 2005
2. *Annual Prize of the Cuban Academy of Sciences* for 2008
2. *Annual Prize of the Cuban Academy of Sciences* for 2004.
3. *Annual Prize of the Cuban Academy of Sciences* for 1998.
4. *Relevant Work of the Cuban Academy of Science* for 1983
5. Elected *Senior Associate-Member of the ICTP* for the period 1997-2003.
6. Elected *Regular Associate-Member of the ICTP* for the period 1991-1997.
7. Appointed *Associate Member* of the Centro de Física y Matemática de América del Sur (CEFIMAS) (Buenos Aires, Argentina 2005)
8. Appointed to a *World Laboratory Fellowship* at Theory Division of CERN (1992)

Biographical Sketch

I born in Havana, Cuba, the 19th of September of 1947 and coursed the secondary studies from 1960 to 1965, and the undergraduate studies at the Faculty of Physics of Havana University, in the period 1965 to 1970.

* After finishing the undergraduate studies, from 1970 to 1976, I started working as a

young professor in the Faculty of Physics of Havana University, doing experimental research on Paramagnetic Resonance (RPE). The results of these activities were the content of my MSc dissertation, defended in 1976 at the same Faculty.

* During 1977, I moved to a new Group of Theoretical Physics at the Cuban Academy. This change realized my longstanding interest to work in Theoretical Physics. In the period from 1977 to 1983, my research was in the area of Field Theory at finite temperature and density, in presence of external fields. In a first stage (1977-1980) the following tasks were considered: 1) The first development of a covariant generalization of the statistical field theory, 2). The calculation of the thermodynamical potential for Quantum Electrodynamics in external magnetic and electric fields and a construction of a model for the quark deconfining phase transition in Chromodynamics. During the PhD stay at Lebedev Physical Institute in Moscow (1980-1983), a solution for the problem of the quantization of gauge fields interacting with external sources was obtained. These results were included in my PhD Dissertation thesis defended in the Theory Division of the Lebedev Physical Institute at the end of 1982. The PhD work was supervised by Prof. Anatoli E. Shabad a member of the Group of Prof. E.S. Fradkin. The results obtained in these times allowed me to be appointed as a **Senior Researcher (Investigador Titular)** at my Center in Havana (Instituto de Cibernética Matemática y Fisica, ICIMAF) and also to receive the award for **Relevant Work of the Cuban Academy of Sciences** for the year 1983.

* After returning to Havana, during the interval 1983 to 1985, my research was directed to study the validity of the Dirac conjecture for constrained dynamical systems including first as well as second-class constraints, a general proof was found which was published. Additionally, a model for the chiral phase transition in QCD was also worked. In 1983, I was elected as a member of the **Scientific Council** of my Centre: ICIMAF.

* Starting in 1985 (during my participation in the 1985 Condensed Matter Workshop at the International Centre for Theoretical Physics, ICTP), my research activity started including themes laying in the interface between Quantum Field Theory and Condensed Matter Physics. Since that time, the theory of Quantum Hall Effect has been one of my main interests. The results in this field between 1985 up to 1995 were: 1) The clarification of relevance of the so called Chern-Simons action in the description of the effect, and the derivation of a set of effective Maxwell equations predicting Meissner like effects in superlattices. Surface and edge mode waves in QHE planar samples were also obtained , 2) The determination of the spatial dependence of the exact propagator for interacting electrons placed in a magnetic field, 3) The construction of a field theory formulation for QHE from which the Streda formula rigorously follows as the coefficient of the Chern-Simons action. The procedure also implied that the conductivity quantization is exact when the filling factor is fixed (i.e. chemical potential laying within a gap of the spectrum), 4) The development of a functional formulation for composite fermions and a model constructed from this approach, which reproduced the main features of the conductivity vs magnetic field in FQHE. 4) The finding of analytic solutions of the Hartree-Fock equations for interacting 2D-electrons in a magnetic field.

* In 1990 I was appointed as a **Regular Associate Member of the ICTP** for the period 1991-1997, a fact that strongly supported my work in this period. I was also granted with **World Laboratory Fellowship** at the European Center of Nuclear Research (CERN) for six months. Those activities helped extended the range of my international research interactions. In particular at CERN, I discussed an idea about a possible solution of renormalizability vs.

unitarity problem of the massive Yang-Mills fields (perhaps, yet being an unsolved question in Quantum Field Theory). The verification of it, although being a difficult task, is yet under consideration. In addition, during the stay in 1990 at the ICTP, I have been specially benefited from the interactions with Profs. A. Salam, J. Strathdee and S. Randjbar-Daemi in connection with the relevance of the Chern-Simons topological action for the description of the QHE, a theme that was also of their interest at that time.

* Afterwards, being at ICTP in 1994, I have started again my former activity on gauge fields, by proposing a modified perturbation expansion for QCD. This procedure, was able to predict, in the simplest approximation, important physical properties of the QCD vacuum. These were by example, the non-vanishing gluon condensate parameter and its spontaneous generation from the usual non-interacting ground state.

* After 1995, the work on Quantum Hall Effect (QHE) was continued. Relations for the order parameter and the self energy spectra were obtained for the previously found analytic Hartree-Fock solutions. Also, nearly minimal HF energy states were determined as well as their representation as solutions of tight binding models. In 1996, a criterion for the stability of a translation invariant ground state for the FQHE was derived. Its application to the Single Mode Approximation for the excitation spectrum around the Laughlin state, indicated the instability of this wave function within the SMA approximation. Moreover, the investigation of the Bethe-Salpeter equation, within the before constructed functional approach for composite fermions, gave also positive indications of a translation invariance breaking. Further research on this basic problem (although our analysis deviates from the current viewpoint in the theory of FQHE) remains under consideration.

* During the period 1997-1998 I had been visiting the Institute of Physics of the Guanajuato University (IFUG) at Leon, Guanajuato, Mexico. There, I was working as a **Visiting Professor** under a **CONACyT** granted support, in a two years cathedra. A result obtained there, was the determination of a solution for the equations for classical scalar fields coupled with gravity. The solution was investigated as a candidate for showing non-collapsing interior space-times for black holes. The physical plausibility of these solutions can imply alternative models for the distribution of matter inside the black holes. In the absence of a singularity at the origin, they would show the collapsed matter as concentrated near a singular boundary with the exterior space. After taking into account the relevance of the Dilaton scalar fields for string theory and the extreme conditions at the black hole horizons, it would be possible that these solutions could also have physical relevance even for the case of fermionic collapsing matter. These considerations strongly motivates us in continuing this theme, which is currently considered in collaboration with students and other colleagues.

* In 1997, after finishing my position as a Regular Associate Member of the ICTP I was also appointed as **Senior Associate Member of ICTP** for the period 1997-2003. Also, my research work in the previous years received the **1998 Annual Prize for Exact Sciences** of the Cuban Academy of Sciences.

* After returning from Mexico to my permanent institution in Havana in 1999, I continued considering the development of the modified perturbative expansion for QCD The first objective was to determine the particular state of the non-interacting theory being able to produce the gluon propagator modification. Further, the procedure was applied to evaluate the quark masses as determined by the considered gluon condensation. Surprisingly, already in the simplest approximations, and after fixing the strength of the condensation to reproduce

the estimated values of the gluon condensation parameter, results for the light constituent quark masses of one third of the nucleon mass were obtained. Moreover, the dependence of the mass on the current quark masses was such that the addition of the values associated to the known quark constituents in many baryon resonances, well reproduced the masses of such hadrons. This activity extends up to the present. By example, in the year 2005, while visiting the Theory Division at CERN and the ICTP, I was able to argue the gauge parameter independence of the modified PQCD proposed. Moreover, it became possible to eliminate the singularities which appeared in the Feynman diagrams due to the presence of Delta functions in the modified propagators. Finally, initial one loop evaluations of the effective action suggested the possible generation of quark masses in the framework of QCD. The work in this direction is yet one of the main motivations of our research activity.

* In July, I visited the Center of Theoretical Studies of Physical Systems (CTSPS) of the Clark Atlanta University. There, I collaborated with Profs. Daniel Bessis and Carlos Handy in finding conditions for the uniqueness of surface densities of charges determining evoked potentials. At CTSPS I had been appointed as an Affiliate Member of the Center at that time.

* In the year 2000, I visited for two month Prof. Francisco Claro at the Faculty of Physics of the Catholic University of Chile. The aim was to join efforts in searching for the breaking of the translation invariance in the ground state of systems undergoing fractional quantum Hall effects. This is another line of work that motivates us in searching for an alternative description of the FQHE. Recently in this years 2009 and 2010 we also determined a class of exact (although excited ones) eigenstates of the FQHE problem after it is restricted to the lowest Landau level. The analysis also allowed to split the lowest Landau level in a set of an arbitrary odd number of subspaces which are left invariant by the Coulomb interaction operator. We expect that this result could bring information about the exact ground state of the FQHE problem. The work in QHE allowed me to receive the **Annual Prize of the Cuban Academy of Science** for Exact Sciences for the year 2004. In addition, in the year 2005, I was also granted with a main recognition to meritory research activity in my country, the **Carlos J. Finlay Medal.**

* In the years 2008, and thanks to starting collaboration with Prof. Robert Brandenberger (McGill, University), we have visited the Dept. of Phys. of the McGill University and the Perimeter Institute for Theoretical Physics (PI) (Waterloo, Canada).

* In a next visit to PI the year 2009, we were able to advance in the study of the dynamical generation of mass in QCD, thanks to obtaining of a modified functional integral for massless QCD in which the effects of the quark condensates are now incorporated through an additional condensate dependent vertex. This fact allowed to evaluate a two loop correction to the effective potential of the condensate which indicates its dynamical generation. The activity on QCD, received the **Annual Prize of the Cuban Academy of Science** for the year 2008.

As mentioned before, the extension of these results is my main current research interest.

* Finally, I would like to conclude by expressing my deep acknowledgement to the enormous support received along many years by me from the Abdus Salam International Centre for Theoretical Physics (Trieste, Italy) and its Associateship and other Programmes.

PROFESSIONAL APPOINTMENTS

1. Faculty of Physics,
   Havana University From 1970 to 1976, Researcher and Professor
2. Cuban Academy of Science,

	Havana	From 1977 to 1982,	Researcher
3.	ICIMAF, Havana	From 1983 to 2006,	Titular Researcher
4.	Institute of Physics, Guanajuato University	From 1996 to 1998,	Cátedra Patrimonial of CONACyT, México.
5.	CERN Theory Division	Six Months in 1992,	World Laboratory Fellowship
6.	Facultad de Física Pontificia Universidad Católica de Chile	August 2001 up to February 2002	Guest Professor

HONORS RELEVANT TO SUPERCONDUCTIVITY RESEARCH:

The Diploma Graduate work of my student, Alejandro Cabo Bizet, defended at the Faculty of Physics in Havana University in the year 2008, received the ***Grand Prize*** of the ***General Student Research Forum*** of the whole Havana University: The work, basically represented *La2CuO4*, conventionally defined as Mott insulator as a Slater (First Principles) insulator. A clear description of the nature of the AF order in superconducting materials also emerged from the analysis, as well as a promising picture for the description of the normal order properties in these materials.

PUBLICATIONS DURING LAST THREE YEARS

1. *Exploring effects of strong interactions in enhancing masses of dynamical origin*
   A. Cabo Montes de Oca
   arXiv:1006.3966v1 [hep-ph] 20 Jun 2010
2. *Hartee Fock Symmetry Breaking Effects in La2CuO4: Hints for connecting the Mott and Slater Pictures and Pseudogap Prediction*
   A. Cabo-Bizet 1, A.Cabo Montes de Oca
   Symmetry 2, 388-417, (2010).
3. *Exact excited eigenstates of the LLL projected FQHE problem*
   A. Cabo and F. Claro
   LANL Preprint Archive: arXiv: 0907.2472 (2009) ,
   submitted to *Physica B*
4. *A path integral formula for quark condensate states in a modified PQCD*
   A. Cabo Montes de Oca, N. G. Cabo-Bizet and A. Cabo-Bizet
   European Physics Journal C,
   Eur. Phys. J. 64: 133–145 *C* (2009)
5. *Could fermion masses play a role in the stabilization of the dilaton in cosmology?*
   Alejandro Cabo and Robert Brandenberger,
   Journal of Cosmology and Astroparticle Physics 02 (2009) 015.
6. *Spontaneous symmetry breaking approach to La2CuO4 properties: Hints for matching the Mott and Slater pictures*
   Alejandro Cabo-Bizet and Alejandro Cabo Montes de Oca
   Physics Letters A 373 (2009) 1865–1869
7. The *stationary density matrix of a pumped polariton system,*

Carlos Andrés Vera, Alejandro Cabo, and Augusto González,
Phys. Rev. Lett. 102 (2009) 126404

8. *Effective Maxwell equations for finite planar samples in QHE regime*
   Alejandro Cabo, Daniel G. Barci, Maurice Oliva
   Physics Letters A 372, 162-175 (2008).
9. *Vacuum polarization of massive spinor fields in static black-string backgrounds*
   Owen Pavel Piedra and Alejandro Cabo de Oca

The renormalized mean value of the quantum Lagrangian and the corresponding components of the energy-momentum tensor for massive spinor fields coupled to an arbitrary gravitational field configuration having cylindrical symmetry are analytically evaluated using the Schwinger-DeWitt approximation, up to second order in the inverse mass value. The general results are employed to explicitly derive compact analytical expressions for the quantum mean Lagrangian and Energy-Momentum tensor in the particular background of the black-string spacetime.

Phys. Rev. D 77, 024044 (2008)

10. *Proposed definitions of the correlation energy density from a Hartree-Fock starting point:*
    The two-electron Moshinsky model atom as an exactly solvable model
    N. H. March, A. Cabo, F. Claro, and G. G. N. Angilella
    Phys. Rev. A 77, 042504 (2008)
11. *Effect of Laughlin correlations on crystalline mean field solutions for the 2D electron gas in a strong magnetic field*
    Alejandro Cabo, Francisco Claro y Danny Martínez-Pedrera,
    ICTP Preprint IC/2007/092 (2007).
12. *Renormalized Wick expansion for a modified PQCD*
    Alejandro Cabo,
    Eur. Phys. J. C 55, 85-93 (2008), *ICTP Preprint* IC/2007/083 (2007).
13. *Static Universe model existing due to the Matter-Dark Energy coupling*
    Alejandro Cabo-Bizet y Alejandro Cabo Monteas de Oca
    ICTP Preprint IC/2007/094 (2007).
14. *On the statistical physics of metastable and non-equilibrium stationary status*
    Alejandro Cabo y Sergio Curilef,
    ICTP Preprint IC/2007/093 (2007).
15. *Hole like analytic mean-field solutions in the fractional quantum Hall regime*
    Alejandro Cabo, Francisco Claro, Alejandro Pérez, y Jerónimo Maze
    Phys. Rev. B 76, 075308 (2007)
16. *Quantization of massive scalar field*
    Owen Pavel Fernández Piedra and Alejandro Cabo Montes de Oca
    Phys. Rev. D 75, 107501 (2007)
17. *Photon Emission as a Source of Coherent Behavior of Polaritons*
    Herbert Vinck-Posada, Boris A. Rodríguez, P. S. Guimaraes, Alejandro Cabo, and Augusto González

We show that the combined effect of photon emission and Coulomb interactions may drive an exciton-polariton system towards a dynamical coherent state, even without phonon thermalization or any other relaxation mechanism. Exact diagonalization results for a finite system (a multilevel quantum dot interacting with the lowest-energy photon mode of a microcavity) are presented in support of this statement.

Phys. Rev. Lett. 98, 167405 (2007)

OTHER CONTRIBUTIONS OF INTEREST TO SUPERCONDUCTIVITY RESEARCH: PROJECTS, GRANTS ETC.

A. The following two works were the content of the Diploma Graduate Dissertion of Alejandro Cabo Bizet at the Faculty of Physics in Havana University. The work received the First Prize in the General Student Research Forum of the Havana University .

1. *Hartee Fock Symmetry Breaking Effects in La2CuO4: Hints for connecting the Mott and Slater Pictures and Pseudogap Prediction,* A. Cabo-Bizet 1, A.Cabo Montes de Oca *Symmetry 2*, 388-417, (2010).
2. *Spontaneous symmetry breaking approach to La2CuO4 properties: Hints for matching the Mott and Slater pictures,* A. Cabo-Bizet and Alejandro Cabo Montes de Oca, *Physics Letters A 373* (2009) 1865–1869

B. The following work also constituted the content of the Diploma Graduation Thesis of Yosdanys Vazquez

3. *About the role of 2D screening in High Temperature Superconductivity,* Y. Vazquez-Ponce, D. Oliva-Aguero and Alejandro Cabo, *Phys. Lett. A* 353, 255-268 (2006)

C. The following set of works, has a close relation with superconductivity: They attempt to explore an old idea of Harald Fritszch: the existence of close links between the spectrum of masses in Particle Physics and the spectra of Superconducting systems which could be hidden in the quantization of massless QCD.

4. *Exploring effects of strong interactions in enhancing masses of dynamical origin* A. Cabo Montes de Oca, arXiv:1006.3966v1 [hep-ph] 20 Jun 2010
5. *A path integral formula for quark condensate states in a modified PQCD* A. Cabo Montes de Oca, N. G. Cabo-Bizet and A. Cabo-Bizet, *European Physics Journal C* 64: 133–145 *C* (2009)
6. *Renormalized Wick expansion for a modified PQCD* Alejandro Cabo, *Eur. Phys. J. C 55*, 85-93 (2008), *ICTP Preprint* IC/2007/083 (2007).
7. *Effective potential evaluations in a modified PQCD.* Alejandro Cabo and Danny Martinez-Pedrera, *Eur. Phys. J. C 47*, 355–364 (2006).

8. *Gauge invariance properties and singularity cancellations in a modified PQCD* Alejandro Cabo and Marcos Rigol, *Eur. Phys. J. C* 47, 95–112 (2006)
9. *Connections between quark masses and quark condensates in a modified perturbative QCD,* A. Cabo, *Jour. of High Energy Phys. 04* (2003) 044. *Phys. Rev. B65* (2002) 2053XX-1
10. *Constituent quark masses from a modified perturbative QCD,* A. Cabo and M. Rigol, *Eur. Phys. Jour. C 23* (2002) 289-300. arXiv:hep-ph/0008003
11. *Baryonic Resonances Mass Spectrum from a Modified Perturbative QCD,* A. Cabo and M. Rigol, arXiv: hep-ph/0007035, 4 July 2000.
12. *Modified initial state for Perturbative QCD*, Marcos Rigol Madrazo, Alejandro Cabo Montes de Oca, *Phys. Rev. D62,* 074018 (2000), Preprint ICTP IC/99/125 (1999).

In: Superconductivity and Condensed Matter …
Editor: Collin J. Eckles
ISBN 978-1-61209-785-5

Chapter 5

BIOGRAPHICAL SKETCH – Superconductivity Research

NAME:	TITLE:	DATE OF BIRTH:
ORION GJON CIFTJA	ASSOCIATE PROFESSOR	NOVEMBER 27, 1967

EDUCATION

Dr. Ciftja was born in Shkoder, Albania. Encouraged by his parents who had completed their University Degrees, respectively, in Physics and Mathematics, he took an early interest in these disciplines. He distinguished himself in these subjects since and early age and eventually became a Physicist.

His University-level education is as follows:

Institution and Location	Degree	Year Conferred	Scientific Field
University of Tirana, Tirana, Albania	Diploma	1991	Physics
International Centre for Theoretical Physics (ICTP), Trieste, Italy	Diploma ICTP	1994	Theoretical Condensed Matter Physics
International School of Advanced Studies (SISSA/ISAS), Trieste, Italy	Master Phil.	1995	Theoretical Condensed Matter Physics
International School of Advanced Studies (SISSA/ISAS), Trieste, Italy	Ph.D.	1997	Theoretical Condensed Matter Physics

Institution and Location	Degree	Year Conferred	Scientific Field
International Centre for Theoretical Physics (ICTP), Trieste I-34100, Italy	Guest Scientist	October 1, 1994- December 31, 1994	Theoretical Condensed Matter Physics
Ames Laboratory, Iowa State University, Ames, Iowa 50011, USA	Post-Doctoral Fellow	1997-1999	Theoretical Condensed Matter Physics
Department of Physics, Texas A&M University, College Station, Texas 77843, USA	Visiting Assistant Professor	1999-2000	Theoretical Condensed Matter Physics
Department of Physics & Astronomy, University of Missouri, Columbia, Missouri 65211, USA	Post-Doctoral Fellow	2000-2002	Theoretical Condensed Matter Physics

CONTACT POINTS

Department of Physics, Prairie View A&M University, Prairie View, TX 77446, USA

RESEARCH AND PROFESSIONAL EXPERIENCE

PROFESSIONAL APPOINTMENTS

- Associate Professor (with tenure), Department of Physics, Prairie View A&M University, Prairie View, Texas 77446, USA (2008-present).
- Assistant Professor, Department of Physics, Prairie View A&M University, Prairie View, Texas 77446, USA (2002-2008).
- Visiting Assistant Professor, Department of Physics, Texas A&M University, College Station, Texas 77843, USA (1999-2000).
- Assistant Professor, Department of Physics, University of Tirana, Albania (1991-1993).

HONORS RELEVANT TO CONDENSED MATTER PHYSICS RESEARCH AND TEACHING

- Scholar Award from the Kavli Institute for Theoretical Physics (KITP), University of California – Santa Barbara for the period 2007-2009.
- Recipient of the third Texas A&M University System (TAMUS) Teaching Excellence Award (Fall 2009) ranked in the Top 5% Category.
- Recipient of the second Texas A&M University System (TAMUS) Teaching Excellence Award (Spring 2009) ranked in the Top 5% Category.
- Recipient of the first Texas A&M University System (TAMUS) Teaching Excellence Award (Fall 2008) ranked in the Top 5% Category.
- Outstanding Faculty Award for Academic Year 2007-2008, Department of Physics, Prairie View A&M University, Prairie View, Texas, USA.

PUBLICATIONS DURING LAST THREE YEARS

1. O. Ciftja, *Anisotropy of quantum Hall phases at filling factor 9/2*, J. Appl. Phys. 107, 09C504 (2010).
2. O. Ciftja and J. Quintanilla, *Effective interaction potentials in the uppermost Landau level,* J. Low. Temp. Phys. 159, 189 (2010).
3. O. Ciftja, Coulomb self-energy and electrostatic potential of a uniformly charged square in two dimensions, Physics Letters A, 374 (7), 981 (2010).
4. O. Ciftja, *A Jastrow correlation factor for two-dimensional parabolic quantum dots,* Modern Physics Letters B, 23 (26), 3055 (2009).
5. O. Ciftja, *Exact results for systems of electrons in the fractional quantum Hall regime II,* Physica B 404, 2244 (2009).
6. O. Ciftja, A. Babineaux, and N. Hafeez, *The electrostatic potential of a uniformly charged ring*, Eur. J. Phys. 30, 623 (2009).
7. O. Ciftja, *Classical behavior of few-electron parabolic quantum dots*, Physica B 404, 2629 (2009).
8. O. Ciftja, *Lamellar-like structures in ferrofluids placed in strong magnetic fields*, Solid State Communications 149, 532 (2009).
9. O. Ciftja, *Exact results for systems of electrons in the fractional quantum Hall regime,* Physica B 404, 227 (2009).
10. O. Ciftja, *Few-electrons semiconductor quantum dots in magnetic field*, Physica Status Solidi (c) 6, 825 (2009).
11. O. Ciftja and C. Wexler, *Anisotropy in two-dimensional electronic quantum Hall systems at half filling of valence Landau levels*, Physica B 403, 1511 (2008).

OTHER CONTRIBUTIONS OF INTEREST TO CONDENSED MATTER PHYSICS RESEARCH: PROJECTS, GRANTS ETC.

My contributions to condensed matter physics research involve typically the areas of strongly correlated electronic systems, quantum Hall effect, nanoscale electronic systems and molecular magnets. Overall, I have published more than forty peer-reviewed papers in these fields, have co-authored a book chapter (Quantum Dots, Nova Science Publishers, 2008) and have been book reviewer of two textbooks (Sears and Zemansky 2004, Rex 2007).

My scholarly achievements in the context of a non-major-research institution have also received national recognition and exposure as illustrated by my selection for the 2007-2009 Scholar Award by the internationally renowned Kavli Institute for Theoretical Physics (KITP) at the University of California, Santa Barbara where I was one of only seven scientists from around the United States to be awarded the KITP Scholar award that year. My research on strongly correlated electronic systems is currently sponsored by the National Science Foundation and my scholarly work in the field has also been featured in few books and publications, most recently the book "Composite Fermions" by Jainendra Jain (Cambridge University Press, 2007, ISBN # 0-521-86232-9).

My scholarly service at national and international level extends in several directions. I serve with the American Physical Society (APS) as Reviewer/Referee for Physical Review Letters, Physical Review B and Physical Review E. With the Institute of Physics (IOP), I serve as Reviewer/Referee for Journal of Physics B, Journal of Physics D and Journal of Physics: Condensed Matter. I also serve as Reviewer of other scholarly journals such as Physics Letters A, Optics Communications, Physica Status Solidi (b), etc. In the past few years, I have served several times as Panel Reviewer for the National Science Foundation (NSF). I also serve as Editorial Board Member for "Applications and Applied Mathematics (AAM)" Journal. Recently, I became Member of the Anacapa Society (which promotes, supports, trains and assists the careers of researchers in theoretical and computational physics at primarily undergraduate institutions) founded at the University of California, Santa Barbara where I have been a summer visitor (KITP Research Scholar) for the last three years. At the local and state level I serve as member of many departmental, college and university committees, I serve as Vertical team member for the College Readiness Assignment Design Team for the College and Career Readiness Initiative of the state of Texas (since April 2009) as well as Sigma and SPS Advisor of student organizations.

AREAS OF RESEARCH INTERESTS

My areas of research interest encompass strongly correlated electronic phenomena, the fractional quantum Hall effect, computational physics, semiconductor quantum dots and nanoscale systems:

- **Strongly correlated electrons and the fractional quantum Hall effect:** The experimental realization of almost ideal two-dimensional (2D) electronic systems opened new frontiers in condensed matter physics and resulted in many discoveries such as the fractional quantum Hall effect (FQHE). It is now believed that the FQHE represents a novel state of matter that arises solely from strong electronic correlations. When the upper Landau levels are half filled it was believed that a Fermi liquid-like state is stable. Recent experiments around half filling of the upper Landau levels have shown surprisingly very strong magneto-transport anisotropies below a critical temperature of about 100 mK. Rather than adopt the more conventional charge density wave approach, one can view the onset of anisotropy as signature of a phase transition to an anisotropic electronic liquid crystalline state with broken rotational symmetry. The formulation of a general theory of liquid crystalline electronic systems and the understanding of the origin the anisotropy is a subject in which I have worked the last few years.

- **Semiconductor quantum dots:** Quantum dots are semiconductor devices consisting of electrons confined in a small region of space, usually in two dimensions. The interplay between quantum confinement, magnetic field and charging effects manifests itself in many complex physical phenomena. The majority of current studies have thoroughly investigated the interplay between confinement, magnetic field and electronic correlations, but less work has been done to understand the impact of those factors on electron's spin correlations. The possibility to use electron's spin as a quantum bit in quantum computing has opened new frontiers in the field. One of the major roadblocks to progress is the absence of a reliable spin readout (control) mechanism in small single or laterally coupled quantum dots. Theoretical work on reliable spin readout mechanism is of major research interest.
- **Molecular magnetism:** With the successful synthesis of a plethora of complex magnetic organic molecules, coordination chemistry has brought forth a new class of nanometer-size magnetic materials. To date, chemists have successfully synthesized magnetic molecules in which are embedded as few as two and as many as thirty interacting magnetic ions. These systems are distinctive because the magnetic interactions can be modified in a controlled way or by variation of magnetic ion species determining the magnitude of the individual spin moments. In a controlled manner one can thus explore the range from (small) quantum spins to essentially large (classical) spins. Theoretical studies of such nanoscale systems in terms of spin models and the understanding various spin effects is a subject that has attracted my interest.

GRANTS AND FUNDED PROJECTS

- Anisotropy in Correlated Electronic Systems in Quantum Hall Regime, National Science Foundation, $138,000, September 15, 2008 – August 31, 2011 (PI).
- Acquisition of a dilution refrigerator with tunnel diode system, National Science Foundation, $321,173, October 1, 2006-September 30, 2009 (Co-PI).
- IGERT: New mathematical tools for next generation materials [Leading Institution: TAMU; Participating Institutions: PVAMU and Texas State University], National Science Foundation, $2,800,000, June 1, 2006-May 31, 2011 (sub-award PI).
- Research on the actinides and related materials at extreme conditions, Department of Energy, $1,500,000, October 1, 2005-September 30, 2007 (Co-PI).
- Physics of nanotechnology at PVAMU, Research Enhancement Program, $3500, September 1, 2003- August 31, 2004 (PI).

ISBN 978-1-61209-785-5

Chapter 6

BIOGRAPHICAL SKETCH – Superconductivity Research

NAME:	TITLE:	DATE OF BIRTH:
TANMOY DAS	POST DOCTORAL RESEARCH SCIENTIST AT LOS ALAMOS NATIONAL LABORATORY, USA	11-20-1981

EDUCATION

Institution and Location	Degree	Year Conferred	Scientific Field
University of Calcutta, India	Bachelor of Science (BSc)	2002	Physics
Indian Institute of Technology (IIT)-Bombay, India	Master of Science (MSc)	2004	Physics
Northeastern University, USA	Doctor of Philosophy (PhD)	2009	Theoretical Physics and Condensed Matter and Materials Science.
Los Alamos National Laboratory, USA	Post Doctoral Research Fellow	Present	Theoretical Physics and Condensed Matter and Materials Science.

CONTACT POINTS

Tanmoy Das
T-4, B262
Los Alamos National Laboratory
Los Alamos, New Mexico 87545

Phone: 505-667-5421
Fax: 505-665-4063
Email: tnmydas@gmail.com
website: www.tanmoydas.com

RESEARCH AND PROFESSIONAL EXPERIENCE

RESEARCH

Los Alamos National Laboratory (2010 -):

My principle research focus is divided into several fields of great current interests.

1) I am presently studying the electronic structure and strong correlation effects of heavy fermionic superconductors including $CeCoIn_5$ series, $PuCOGa_5$ series and URh_2Si_2 series. These compounds contain highly correlated *f*-electrons, sometimes accompanies by *p*- and *d*-orbitals. Thus understanding the mechanism of electron pairing or superconductivity in these materials is highly challenging than in normal metallic superconductors. My research is directed in this direction to untangle both the normal and superconducting ground state properties using tight-binding band structure in the presence of strong electron-electron correlations. I will also study the effect of disorder and impurity to the suppression of superconducting transition temperature in these class of materials.
2) My second main research interest is to study various high-energy physical theories as applicable to the newly discovered topological insulators. These materials are also shown experimentally to become superconductors in the pristine phase, or with external perturbations.
3) My research interests also extend to the high-temperature superconductors in the family of copper-oxide and iron-arsenic compounds. Here, I am continuing my research on understanding the spectroscopic properties of these materials including angle-resolved photoemission spectroscopies (ARPES), scanning tunneling microscopy (STM) and inelastic neutron scattering (INS) experiments.

Doctor of Philosophy (2004-2009) at Northeastern University, Boston, USA:

I mainly focused on developing an intermediate coupling model of cuprates to incorporate fluctuations in a mean-field model of pseudogap and superconductivity competition. The model is called quasiparticle-GW (QP-GW) model. The model consistently describes a variety of spectroscopies including angle-resolved photoemission spectroscopy (both Fermi surface topology, low and high-energy kinks), optical spectra, penetration depth measurements, specific heat measurements, scanning tunneling microscopy, magnetic resonance behavior, x-ray absorption spectra, resonant inelastic x-ray scatterings and other light scattering spectroscopies. All results consistently reflect the interplay between the magnetism and superconductivity and that the intermediate coupling strength (U~bandwidth) is more appropriate for cuprates. The obtained results are compared with quantum Monte

Carlo (QMC) and other variational cluster approaches and contrasted with other strong coupling calculation such as infinite-U Hubbard model or t-J models.

PROFESSIONAL APPOINTMENTS

2010 –Adjunct Research Assistant Professor in Physics at Northeastern University, Boston, USA.
2010 – Post Doctoral Research Scientist at Los Alamos National Laboratory, Los Alamos, USA.
2009 – 2010 Post Doctoral Research Fellow at Northeastern University, Boston, USA.

REFEREE OF JOURNALS

Physical Review Letters
Physical Review B
Journal Of Physics And Chemistry Of Solids

HONORS RELEVANT TO SUPERCONDUCTIVITY RESEARCH:

2010 - Invited in Marquis' Whos'Who biographical magazine.
2005 - Current Research Assistantship scholarship under DOE grant with A. Bansil PI.
2004 - Nominated for best M.Sc. project in IIT-Bombay, India.
2004 - Qualified National Eligibility Test (NET), India.
2003 & 2004 Selected for Summer Fellowship S.N. Bose National Center for Basic Science, India.

EXPERTISE IN NUMERICALL TECHNIQUES AND COMPUTATIONS SKILLS

1. I have learned the professional first principle code TB-LMTO (Tight Binding Linear Muffin-Tin Orbital code from the German group of Prof. O. K. Andersen) in depth.
2. For the purpose of extracting best tight binding parameters which can describe the basic LDA band, have learnt the first principle codes of Wien-2K (commercial), KKR (Internal code of Bansil's group).
3. I have more than 9 years of extensive experience in programming in Fortran 77/95 and MPI parallelization techniques, Matlab, Mathematica and C/C++.
4. I have more than 9 years of extensive experience in programming in Fortran 77/95 and MPI parallelization techniques, Paraview, Matlab, Mathematica and C/C++.

PROFESSIONAL ACTIVITIES AND ASSOCIATIONS

2008 - Invited member of Science Magazine, maintained and sponsored by American. Association for the Advancement of Science (AAAS) based on nominations.
2009 - Member of Computations Materials Science Network (CMSN) , sponsored by the U.S. Department of Energy (DOE), Office of Basic Energy Sciences.
2010 - Active member of American Physical Society (APS) and Nature Publishing Group.

PUBLICATIONS DURING LAST THREE YEARS
(2007-2010)

1. J. Nieminen, Ilpo Suominen, Tanmoy Das, R. S. Markiewicz, and A. Bansil, *Doping dependence of STM spectra and the role of strong correlation and van-Hove singularity in Bi2212,*Physical Review Letters (2010).
2. Tanmoy Das, R. S. Markiewicz, and A. Bansil, *Origin of low-energy kink in cuprates: magnon or phonon?* Physical Review Letters (2010).
3. Susmita Basak, Tanmoy Das, Hsin Lin, R. S. Markiewicz, and A. Bansil, *Modification of RIXS spectra of cuprates due to self energy and matrix element effects*, Physical Review Letters (2010).
4. Yung Jui Wang, B. Barbiellini, Hsin Lin, Tanmoy Das, Susmita Basak, P. E. Mijnarends, S. Kaprzyk, R. S. Markiewicz, and A. Bansil, *Dynamical structure factor computations in extended momentum space in electron doped cuprates,* Physical Review B (2010).
5. Towfiq Ahmed, John J. Rehr, Joshua J Kas, Tanmoy Das, Hsin Lin, R.S. Markiewicz, B. Barbiellini, A. Bansil, *Fermi liquid description of XAS in overdoped LSCO*, Physical Review B (2010).
6. Hsin Lin, Tanmoy Das, R.S. Markiewicz, and A. Bansil, *Origin of doping dependence of Hubbard U in cuprates,* Physical Review B (2010).
7. R.S. Markiewicz, Tanmoy Das, and A. Bansil, *Reconciling waterfalls and optical glue in the cuprates*, Physical Review Letters (2010).
8. Tanmoy Das, R. S. Markiewicz, and A. Bansil, *Origin of magnetic resonance spectrum in high-temperature copper-oxide superconductors*, Proceeding of National Academy of Science (PNAS), USA (2010).
9. Tanmoy Das, R. S. Markiewicz, and A. Bansil, *Optical model-solution to the competition between a pseudogap phase and a Mott-gap phase in high-temperature cuprate superconductors*, Physical Review B 81, 174504 (2010).
10. Tanmoy Das, R. S. Markiewicz, and A. Bansil, *Emergent non-Fermi-liquid behavior due to Fermi surface reconstruction in the underdoped cuprate superconductors*, Physical Review B 81, 184515 (2010).
11. Susmita Basak, Tanmoy Das, Hsin Lin, J. Nieminen, M. Lindroos, R.S. Markiewicz, and A. Bansil, *Origin of the high-energy kink or the waterfall effect in the photoemission spectrum of the* $Bi_2Sr_2CaCu_2O_8$ *high-temperature superconductor*, Physical Review B 80, 214520 (2009).
12. R. S. Markiewicz, Tanmoy Das, and A. Bansil, *Evidence of anomalous spectral weight transfer and intermediate coupling range in cuprates: correlations across spectroscopies,* Physical Review B (2009).
13. R. S. Markiewicz, Tanmoy Das, Susmita Basak, and A. Bansil, *Intermediate coupling model of cuprates: adding fluctuations to a weak coupling model of pseudogap and*

superconductivity competition, Submitted to Journal of Electron Spectroscopy and Related Phenomena (2010).
14. Tanmoy Das, R.S. Markiewicz, and A. Bansil, *A competing order scenario of two-gap behavior in hole doped cuprates,* Physical Review B 77, 134516 (2008).
15. Tanmoy Das, R.S. Markiewicz, and A. Bansil, *Superconductivity and topological Fermi surface transitions in electron-doped cuprates near optimal doping,* Journal of Physics and Chemistry of Solids, 69, 2963, (2008).
16. Tanmoy Das, R.S. Markiewicz, and A. Bansil, *Nodeless d-wave superconducting pairing due to residual antiferromagnetism in underdoped* $Pr_{2-x}Ce_xCuO_{4-\delta}$, Physical Review Letters 98, 197004 (2007).

OTHER CONTRIBUTIONS OF INTEREST TO SUPERCONDUCTIVITY RESEARCH: PROJECTS, GRANTS ETC.

I am presently serving as an Editor and also contributing a chapter to an Upcoming book on "Theory and Applications of Superconductivities" by Nova Science Publishers.

In: Superconductivity and Condensed Matter ...
Editor: Collin J. Eckles
ISBN 978-1-61209-785-5

Chapter 7

BIOGRAPHICAL SKETCH – Superconductivity Research

NAME:	TITLE:	DATE OF BIRTH:
MOKHOTJWA SIMON DHLAMINI	DR.	13/12/1977

EDUCATION

Institution and Location	Degree	Year Conferred	Scientific Field
University of the North, RSA	BSc.	2001	Physics and Mathematics
University of the Free State, RSA	BSc. Hons.	2002	Physics
University of the Free State, RSA	MSc.	2004	Physics
University of the Free State, RSA	PhD	2008	Physics
Council for Scientific and Industrial Research (CSIR)	Post Doctoral Fellow	2008	Nano-Structured Materials: Nano-phosphors

RESEARCH AND PROFESSIONAL EXPERIENCE

PROFESSIONAL APPOINTMENTS

FULL-TIME

2010 Jul.; Senior Research Scientist, Council for Scientific and Industrial Research (CSIR), RSA.

Part-time

2008 Oct. – 2010 June; Post-Doctoral Fellow, CSIR, RSA.

2008 (Feb. – Dec.); Laboratory technician and Moderator of first year Physics examination paper at Central University of Technology, Bloemfontein, RSA.

2005 (May) – 2008 (Dec.); Post-Graduate Intern student at National Metrology Institute of South Africa (NMISA).

2002 (Feb.) – 2008 (Sep.); Laboratory assistant at the University of the Free State.

2000 – 2001; Student assistant in the first year Mathematics at the University of the North.

Honors relevant to research

- GOODFELLOW award for the most outstanding PhD work in the form of paper in 2009; Condensed Matter Physics and Materials Science group; South African Institute of Physics (SAIP) 54th conference
- Best PhD presentation in 2007; Condensed Matter Physics and Materials Science group; South African Institute of Physics (SAIP) 52nd conference

Adjudicator: Nanoscience Young Researcher Symposium (NYRS), MSc and PhD students from different universities around Gauteng province which was held at CSIR in 2009.

Publications during last three years

1. M.S. Dhlamini, G.H. Mhlongo, and K.T. Hillie, Effects of accelerating beam voltage on the luminescence of Er^{3+} and Tm^{3+} doped SiO_2 phosphor prepared by sol-gel process, *Journal of Optical Materials*, Accepted, 2010.
2. G.H. Mhlongo, O.M. Ntwaeaborwa, M.S. Dhlamini, H.C. Swart, K.T. Hillie, Cathodoluminescence properties of $SiO_2:Pr^{3+}$ and $ZnO:SiO_2:Pr^{3+}$ phosphor nanopowders, *Journal of Material Science*, DOI 10.1007/s10853-010-4563-8.
3. H.C. Swart, J.J. Terblans, E. Coetsee, V. Kumar, O.M. Ntwaeaborwa, M.S. Dhlamini, and J.J. Dolo, Auger Electron Spectroscopy and X-ray Photoelectron Spectroscopy study of the electron stimulated surface chemical reaction mechanism for phosphor degradation, *Surface and Interface Analysis*, 42 (2010) 922 – 926.
4. M.S. Dhlamini, O.M. Ntwaeaborwa, H.C. Swart, J.M. Ngaruiya and K.T. Hillie, Sensitized luminescence through nanoscopic effects of ZnO encapsulated in $SiO_2:Tb^{3+}$ sol gel derived phosphor, *Physica B,* 404 (2009) 4406–4410.
5. H.C. Swart, J.J. Terblans, O.M. Ntwaeaborwa, E Coetzee, B.M. Mothudi and M.S. Dhlamini, Photon emission mechanisms of different phosphors, Nucl. Inst. and methods in physics research section B: Beam interactions with materials and atoms, 267 (2009) 2630 – 2633.
6. O.M. Ntwaeaborwa, M.S. Dhlamini, J.J. Terblans, H.C. Swart, The effects of phonon-mediated energy transfer from ZnO nanoparticles to SiO_2:PbS Book of extended abstracts - 14th International Workshop on Inorganic and Organic Electroluminescence & 2008 International Conference on the Science and Technology of Emissive Displays and Lighting, 9-12 September 2008, Italy (Rome) pp 177 – 179 (ISBN 88-8286-194-5) .

7. M.S. Dhlamini, J.J. Terblans, O.M. Ntwaeaborwa , J.M. Ngaruiya , K.T. Hillie , J.R. Botha and H.C. Swart, Photoluminescence properties of powder and pulsed laser deposited PbS nanoparticles in SiO_2, *Journal of Luminescence*, 128 (2008) 1997 – 2003.
8. M.S. Dhlamini, J.J. Terblans, R.E. Kroon, O.M. Ntwaeaborwa, J.M. Ngaruiya, J.R. Botha and H.C. Swart, Photoluminescence properties of SiO_2 surface passivated PbS nanoparticles, *South African Journal of Science*; *Research Letters*, 104 (2008) 398 – 400.
9. J.J. Dolo, H.C. Swart, J.J. Terblans, E. Coetzee, M.S. Dhlamini, O.M. Ntwaeaborwa and B.F. Dejene, Degradation of commercial Gd_2O_2S phosphor, *Phys. Stat. Sol., (c),* 5(2) (2008) 594 – 597.
10. O.M. Ntwaeaborwa, M.S. Dhlamini, R. Harris, J.R. Botha, U. Buttner and H.C. Swart, Characterization of sol-gel SiO_2:Ce,Tb powder and pulsed laser deposited thin film phosphors, *Phys. Stat. Sol., (c),* 5(2) (2008) 602 – 605.
11. M.S. Dhlamini, J.J. Terbalns, O.M. Ntwaeaborwa and H.C. Swart, Preparations and luminescent properties of PbS nanoparticle phosphors incorporated in a SiO_2 matrix , *Phys. Stat. Sol., (c),* 5(2) (2008) 598 – 601.
12. M.S. Dhlamini, J.J. Terbalns, O.M. Ntwaeaborwa and H.C. Swart, Synthesis and degradation of PbS nanoparticles embedded in SiO_2, (SiO_2:PbS), *Surface Review and Letters*, 14(4) (2007) 697 – 701.
13. Swart H.C., Terblans J.J., Coetzee E., Ntwaeaborwa O.M., Dhlamini M.S. and Holloway P.H., Review on electron stimulated surface chemical reaction (ESSCR) mechanism for phosphor degradation, *American J. Vac. Sci. Technol.,* A25(4) (2007) 917 – 921.

OTHER CONTRIBUTIONS OF INTEREST TO RESEARCH: PROJECTS, GRANTS ETC.

- Preparation and characterization of up- and down-conversion nanophosphors: Possible applications; biological labeling and imaging, silicon solar cells, display technology and gas sensors.

In: Superconductivity and Condensed Matter …
Editor: Collin J. Eckles
ISBN 978-1-61209-785-5

Chapter 8

BIOGRAPHICAL SKETCH – Superconductivity Research

NAME:	TITLE:	DATE OF BIRTH:
GLADKOV	PROFESSOR	AUGUST 19, 1951

EDUCATION

Institution and Location	Degree	Year Conferred	Scientific Field
Moscow State University Moscow, Lenin's Hill, 1 Physical Problem Institute Academy of Sciences of the USSR, Moscow, Kosygin's Str. 2. Chemical Physics	Magister Phil. Dr. Doctor of Physical and	1973 1981	Physic of Solid State Theoretical Physics
Institute Academy of the Sciences of the USSR. Moscow, Kosygin's Str. 4.	And Mathematic al Sciences	1995	Solid State Physics

CONTACT POINTS

Academica Anohina str. 38 – 1, apt. 9. 119602, Moscow, Russia.
Tel.: 8(405)434-75-05, 8(917)574-69-97
E-mail: Sglad@newmail.ru

RESEARCH AND PROFESSIONAL EXPERIENCE

2010/present. Moscow Aviaition Institute (Technical University), Professor of Mathematical Modeling Chair. Read lectures on the following: 1. Variation calculation. 2. Theory of

Probabilities . 3. Quantum mechanics. 4. Statistical Physics. 5.Theory of critical phenomenon. 6. Physics of Nanostructures.

1999 – 2010. Moscow State Regional University, Professor. 1. Mathematical Analysis. 2. Hydrodynamics. 3. Variation Calculation. 4. The Mathematical Physics Equation.

Researches them: Porous structures.

Monographs:

1. S.O. Gladkov. Dielectric properties of porous media. Springer 2003
2. S.O. Gladkov. Collection of problems on theoretical and mathematical physics. Moscow. Publisher Phys. And Math. 2006, second edition 2010.

1981 – 1999. Chemical Physics Institute Academy of the Sciences of the USSR. Senior researcher. The them of researches: 1. Glass – like materials, 2. Porous structures, 3. Magnetic materials, 4. Semiconductors and metals.

Review:

1. S.O. Gladkov. The kinetics of nuclear ordered systems. Physics Reports. 1989. V. 182. N 4,5/ PP. 211- 365.

Monographs:

1. S.O. Gladkov. Physics of porous structures. Moscow . Publisher "Science".1997.
2. S.O. Gladkov. Physics of composites: thermodynamics and dissipative properties. Moscow. Publisher "Science". 1999.

1974 – 1981. Institute of Physical Problems Academy of Sciences of the USSR. The them of researches: Magnetic Phenomenon at ultra low temperatures.

Professional Appointments

Professor

Honors relevant to research

Soros's professor

Publications during last three years

1. S.O. Gladkov. The gas – kinetic model of heat conductivity of heterogeneity substances. Journal of Applied physics. 2008. V. 78. N 7. PP. 12 – 116.

2. S.O. Gladkov. On mathematical description of crystallization process as determination chaos problem. Journal of Applied Physics. 2008. V. 78. N 7. PP. 136 – 140.

3. S.O. Gladkov. Nonlinear dispersion of electromagnetic spectrum. Nonlinear Peace. 2008. V. 6. N1. PP. 23 – 28 (in Russian).

4. S.O. Gladkov, R.G. Rabadanov. On dynamics of thin string in viscous medium. Nonlinear Peace. 2008. V. 6. N 7. PP. 33 – 38 (in Russian).

5. S.O. Gladkov. Theoretical description of the temperature distribution on Earth's atmosphere. Atmospheric and Ocean Optics. 2009. V. 22. N 2. PP. 180 – 185.

6. S.O. Gladkov, S.B. Bogdanova. The heat – transfer theory for quasi – n – dimensional systems. Physica B: Condensed Matter 2010. V. 405. N 8. PP. 1973 – 1975.

OTHER CONTRIBUTIONS OF INTEREST TO RESEARCH: PROJECTS, GRANTS ETC.

Russian Grant Foundation 1999

In: Superconductivity and Condensed Matter ...
Editor: Collin J. Eckles
ISBN 978-1-61209-785-5

Chapter 9

BIOGRAPHICAL SKETCH – Superconductivity Research

NAME:	TITLE:	DATE OF BIRTH:
AJAWAD I. HAIJA	PROFESSOR	12/10/1945

EDUCATION

Institution and Location	Degree	Year Conferred	Scientific Field
The Pennsylvania State University, University Park, State College, PA, U. S. A.	Ph. D.	May 1977	Solid State Physics/Thin Films and Multilayer structures.
The University of Alexandria	B. Sc.	June, 1968	Physics/1st Honor Degree

RESEARCH AND PROFESSIONAL EXPERIENCE

PROFESSIONAL APPOINTMENTS

Aug 2009 - Present: Professor at Indiana University of Pennsylvania, IUP, Indiana, PA.
Aug 2003 – Aug 2009: Associate Professor at Indiana University of Pennsylvania, IUP, Indiana, PA. Tenured effective Aug 28, 05.
Aug 2000 – Aug 2003: Assistant Professor at Indiana University of Pennsylvania, IUP, Indiana, PA
Aug 1st, 1999-Aug 2000. Professor at the University of Jordan, Amman, Jordan
1988 – Aug 1st, 1999: Associate Professor at the University of Jordan, Amman, Jordan
1977 –1988: Assistant Professor at the University of Jordan, Amman, Jordan

PUBLICATIONS DURING LAST THREE YEARS

1. "Effective Characteristic Matrix of Ultra Thin multilayer structures" A. J. Haija, W. Larry Freeman, and Thomas Roarty, Optica Applicata, Vol XXXVI, No 1, (2006).
2. "CEMA Corrections of the Optical Properties of Ultra Thin Ag-Sio Superlattices", A. J. Abu El-Haija, W. Larry Freeman, Angela Musto, and Ekapop Pairam, Vol 9, Issue ¾, 2008.
3. "Computational Criterion For Application of the Characteristic Effective Medium Approximation to Ultrathin Co-Au Multi Bilayer Structures" A. J. Haija, W. Larry Freeman, and Matthew DeNinno, , Physica B, 403, (2008).
4. "Effective Optical Constants And Effective Optical Properties of Ultrathin Trilayer Structures",
   A. J. Haija, W. Larry Freeman, and Rachel Umbel, submitted to Physica B, March 2010.
5. Review of several Chapters of The Physics Textbook: College Physics, By Giambattista, Giambattista, and Richardson, upon a request from the Publisher McGraw Hill, May 04. My name has been recognized and cited as one of the reviewers and focus group attendees in the text's preface, page xxx, published by McGraw-Hill (2007).
6. Member of McGraw Hill ARIS Board of Advisors (BOA) from 2008 - June 2009.
7. Recognized as one of the accuracy reviewers of the text "College Physics" by Giordano. The text has been published by Cengage Learning, Brooks/Cole (2009). The recognition as been cited in the preface of the text.
8. Reviewer of several chapters of the text: "Physics", 8 edition, by Cutnell & Johnson, published by John Wiley. The Publisher has acknowledged my contribution in the text's 8th edition/2009.

In: Superconductivity and Condensed Matter … ISBN 978-1-61209-785-5
Editor: Collin J. Eckles

Chapter 10

BIOGRAPHICAL SKETCH – Superconductivity Research

NAME:	TITLE:	DATE OF BIRTH:
JUNHUA HAO	MR.	1983.07.03

EDUCATION

Institution and Location	Degree	Year Conferred	Scientific Field
Hebei University of Technology, China	Bachelor	2006	Applied Physics
Nankai University, China	Master	2009	Condensed Matter Physics

RESEARCH AND PROFESSIONAL EXPERIENCE

2009.06 to present Ren'ai College of Tianjin University Physics Teacher

Since fall 2009, I have been working in Ren'ai College of Tianjin University and taking the full responsibility for teaching courses of Physics Experiment and Physics. Besides teaching, I have also been doing some research about mechanical properties of doped materials under high pressure using the first-principles method.

2007.09 -- 2009.06 TJU and NKU cooperation Project sponsored by Ministry of Education Core Member

In academic year of this period, I'm a graduate student in Condensed Matter Physics research group, and my research field is calculating the properties of several materials under high pressure using the first principles method within the framework of Density Functional Theory. The peculiar properties of many oxides and mineral materials have a wide range of applications (fuel cell, storage of energy, catalysts, photovoltaic, thermoelectricity…), thus we investigate these materials as well as the composition of the earth's interior and other planetary interiors under high pressure, and then the theoretically forecast configuration and characteristics of the materials can provide theoretical direction and evidence for materials design and preparation.

In the Project, I'm mainly responsible for calculating the properties of semiconductor materials (ZnO, SiO_2) and mineral material CaTe under pressure using the CRYSTAL program.

2008.09 15th National Conference on Condensed Matter Theory and Statistical Physics
Reporter

Attend the meeting and make a post report about the properties of materials under high pressure.

PUBLICATIONS DURING LAST THREE YEARS

1. **Junhua Hao**, Zhiqiang Wu, Zheng Wang, et al. "First principles calculations of structural phase transformation in CaTe at high Pressure" Physica B 404 (20)3671-3673(2009).
2. **Junhua Hao**, Zhiqiang Wu, Zheng Wang, et al."First Principles Calculation of SiO_2 at High Pressure", Chinese Journal of High Pressure Physics 24 (4) 260-266 (2010).
3. **Junhua Hao**, Zhiqiang Wu, Zheng Wang, et al."First-principles Study of Structural Phase Transformation and Electronic Structure of ZnO Under High Pressure", Chinese Journal of Computational Physics 27 (5) 37-42 (2010).
4. **Junhua Hao**, Zhiguang Guo, Qinghua Jin, "First principles calculation of structural phase transformation in Mg_2Si at high pressure" to be submitted Solid State Communications.

In: Superconductivity and Condensed Matter …
Editor: Collin J. Eckles

ISBN 978-1-61209-785-5

Chapter 11

BIOGRAPHICAL SKETCH – Superconductivity Research

NAME:	TITLE:	DATE OF BIRTH:
KONG HUI	ASSOCIATE PROFESSOR	27 Jul 1980

EDUCATION

Institution and Location	Degree	Year Conferred	Scientific Field
Department of Materials Science and Engineering, University of Science and Technology of China	BS	2002	Material Chemistry
Department of Computer Science and Technology, University of Science and Technology of China	Dual Degree	2002	Computer Science and Technology
Department of Materials Science and Engineering, University of Science and Technology of China	Ph.D	2007	Material Physics and Chemistry

RESEARCH AND PROFESSIONAL EXPERIENCE

PROFESSIONAL APPOINTMENTS:

- **March, 2005 –present**

  Investigated the different charge ordering (CO) transitions at low temperature in $La_{1-x}Sr_xFeO_3$;

  Systematically studied the ultrasonic sound velocity and attenuation properties in $La_{1-x}Sr_xFeO_3$, $La_{1/3}Sr_{2/3}Fe_{1-x}M_xO_3$ (M=Co, Cr, Mn) and $R_{1/3}Sr_{2/3}FeO_3$(R=La, Pr, Gd). A series of dramatic ultrasonic anomalies are observed in these compounds, which are attributed to the electron-phonon coupling via the Jahn-Teller effect of Fe^{4+} and the breathing-type distortion of Fe-O octahedron.

  These results suggest that the stability of Fe^{4+} and Fe^{5+} plays an important role for the charge-ordered state in $La_{1-x}Sr_xFeO_3$, and reveal the evolution of complex CO phenomena in $La_{1-x}Sr_xFeO_3$;

- **August, 2004 –February, 2005**

  Be a Process Engineer (intern) in the Applied Materials China, Beijing; Mastered the PVD process in the VLSI;

- **September, 2003 –July, 2004**

  Took part in the design of a simple HTCVD system for SiC films;

PUBLICATIONS DURING LAST THREE YEARS

1. **Kong Hui,** Wang JianJun, Guo Shangxing, Zhou Li and Li Jie, "Copper Removal From Molten Steel With FeS-Na_2S Slag", High Temperature Materials and Processes, 2009, Vol.28, No.1-2, 67-73;;
2. **Kong Hui**, Li Jie, Zhu Changfei, "Ultrasonic signature of Jahn-Teller effect in $La_{1/3}Sr_{2/3}Co_{1-x}Fe_xO_3$ perovskites", Physica B: Physics of Condensed Matter, 2009, Vol.404, 1128-1131;
3. **Kong Hui**, Zhu Changfei, "Charge ordering correlated ultrasonic anomalies in $La_{1/3}Sr_{2/3}Fe_{1-x}Cr_xO_3$ perovskites", Journal of Alloys and Compounds, 2009, Vol.478, 805-808;
4. **Kong Hui**, Zhu Changfei, "Variation of Ultrasonic behaviors in $R_{1/3}Sr_{2/3}FeO_3$ (R=La, Pr, Gd)", Europhysics Letters, 2009, Vol.86, 54001-54005;
5. **Hui Kong**, Changfei Zhu, "An ultrasonic study on complex charge ordering phenomena for $La_{1-x}Sr_xMO_3$", Journal of Physics: Condensed Matter, 2008, Vol.20, 115211-115214

6. Jianxing Yi, **Hui Kong**, Changfei Zhu, "Ultrasonic study of the Jahn-Teller effect in $La_{1-x}Ca_xCoO_3$ ($0.1 \leq x \leq 0.2$)", Journal of Alloys and Compounds, 2009, Vol.474, 38-41;

OTHER CONTRIBUTIONS OF INTEREST TO RESEARCH: PROJECTS, GRANTS ETC.

Led R&D projects in:

1. National Natural Science Foundation of China (Grant No.11004002);
2. The Key Program of Scientific Research Fund of Anhui Provincial Education Department (No. KJ2008A018)

In: Superconductivity and Condensed Matter …
Editor: Collin J. Eckles
ISBN 978-1-61209-785-5

Chapter 12

BIOGRAPHICAL SKETCH – Superconductivity Research

NAME:	TITLE:	DATE OF BIRTH:
SHADIA JAMIL IKHMAYIES	DR	5/13/1960

EDUCATION

Institution and Location	Degree	Year Conferred	Scientific Field
University of Jordan, Jordan	B.S.C	1983	Physics
University of Jordan, Jordan	M.S.C	1987	Molecular Physics
University of Jordan,	Ph.D	2002	Semiconductors, Thin Films, and Solar Cells

RESEARCH AND PROFESSIONAL EXPERIENCE

PROFESSIONAL APPOINTMENTS

2010-now : Assistant professor, Faculty of Science and Information Technology, Al-Isra University.

2004-2009: Assistant professor, Physics Department-Applied Science Private University.

2003-2004: Part time Lecturer, Physics Department-University of Jordan, summer semester, 1st and 2nd semesters then the next summer semester.

2003, 2nd semester: Part time Lecturer, Collage of Technology Engineering-Al-Blqa' Applied University.

1997-2000: Faculty Member (Teacher), Faculty of Basic Science, Al-Isra University.

1987-1997: Labaratory Instructor, Physics Department-University of Jordan.

Publications during Last Three Years

Journal Papers

1. **Shadia J Ikhmayies** and Riyad N Ahmad-Bitar, Optical properties of nanocrytalline CdTe thin films, *Physica B: Condensed Matter*, (accepted).
2. **Shadia J Ikhmayies**, Naseem M. Abu El-Haija and Riyad N. Ahmad-Bitar. Characterization of Undoped Spray-Deposited ZnO Thin Films of Photovoltaic Applications, *FDMP: Fluid Dynamics & Materials Processing* (accepted).
3. **Shadia J Ikhmayies**, Naseem M. Abu El-Haija and Riyad N. Ahmad-Bitar. The Influence of Annealing in Nitrogen Atmosphere on the Electrical, Optical and Structural Properties of Spray- Deposited ZnO Thin Films, *FDMP: Fluid Dynamics & Materials Processing*.(accepted).
4. **Shadia J Ikhmayies** and Riyad N Ahmad-Bitar, A Comparison Between the Electrical and Optical Properties of CdS:In Thin Films for Two Doping Ratios, *Jordan Journal of Mechanical and Industrial Engineering JJMIE* 4(1) (2010): 111-116.
5. **Shadia J Ikhmayies** and Riyad N Ahmad-Bitar, Characterization of the SnO_2:F/CdS:In structures prepared by the Spray pyrolysis technique, *Solar Energy Materials and Solar Cells* 94 (5) (2010): 878-883.
6. **Shadia J Ikhmayies** and Riyad N Ahmad-Bitar, The Influence of the substrate temperature on the photovoltaic properties of spray-deposited CdS:In thin films, *Applied Surface Science* 256 (11) (2010): 3541-3545.
7. **Shadia J Ikhmayies,** Naseem M. Abu El-Haija and Riyad N. Ahmad- Bitar, Electrical and optical properties of ZnO:Al thin film prepared by the spray pyrolysis technique, *Physica Scripta* 81(1) (2010) art. no.015703
8. **Shadia J Ikhmayies** and Riyad N Ahmad-Bitar, Effect of the Substrate Temperature on the Electrical and Structural Properties of Spray-Deposited SnO2:F Thin Films, *Materials Science in Semiconductor Processing* 12(3) (2009): 122-125.
9. Shadia J Ikhmayies and Riyad N Ahmad-Bitar, Effects of processing on the electrical and structural properties of spray-deposited CdS:In Thin Films, *Physica B: Condensed Matter* 404 (16) (2009): 2419-2424.
10. **Shadia J Ikhmayies** and Riyad N Ahmad-Bitar, Effects of annealing in nitrogen atmosphere and HCl-etching on the photoluminescence spectra of spray-deposited CdS:In thin films, *Applied Surface Science* 255 (20) (2009): 8470-8474.
11. **Shadia J Ikhmayies** and Riyad N Ahmad-Bitar, The effects of post-treatments on the photovoltaic properties of spray-deposited SnO_2:F thin films, *Applied Surface Science* 255 (2008): 2627-2631.
12. **Shadia J Ikhmayies** and Riyad N Ahmad-Bitar, Effect of Film thickness on the Electrical and Structural Properties of CdS: In Thin Films, *American Journal of Applied Sciences* 5 (9) (2008): 1141-1143.
13. **Shadia J Ikhmayies** and Riyad N Ahmad-Bitar, Effect of Processing on the Electrical Properties of Spray-Deposited SnO2:F Thin Films, *American Journal of Applied Sciences* 5(6) (2008): 672-677.
14. **Shadia J Ikhmayies** and Riyad N Ahmad-Bitar, Direct observation of Infrared Photoluminescence of Spray-Deposited CdS:In Thin Films, *J. Luminouscence*, 128(4), (2008): 615-619.

CONFERENCE PAPERS (ORAL PRESENTATIONS AND PUBLISHED IN THE PROCEEDINGS):

1. **Shadia J Ikhmayies** and Riyad N Ahmad-Bitar. Production of $CdS_{1-x}Te_x$ Thin Films and Bandgap Investigation of the Produced Solid Solution. *TMS 2011 140th Annual Meeting & Exhibition,* which will be held in San Diego, California, USA from February 27 to March 3, 2011. (abstract submitted)
2. **Shadia J Ikhmayies** and Riyad N Ahmad-Bitar. Photoluminescence of the Interface of SnO_2:F/CdS:In/CdTe Thin Film Solar Cells Prepared Partially by the Spray Pyrolysis Technique. *TMS 2011 140th Annual Meeting & Exhibition,* which will be held in San Diego, California, USA from February 27 to March 3, 2011. (abstract submitted).
3. **Shadia J Ikhmayies** and Riyad N Ahmad-Bitar. An Investigation of the Photoluminescence and Transmittance of $CdS_{1-x}Te_x$ Thin Films. *TMS 2011 140th Annual Meeting & Exhibition,* which will be held in San Diego, California, USA from February 27 to March 3, 2011. (abstract submitted).
4. **Shadia J Ikhmayies** and Riyad N Ahmad-Bitar. An Investigation of the Bandgap and Urbach Tail of Spray-Deposited SnO_2:F Thin Films. 11th Eurasia Conference on Chemical Sciences 6-10 October 2010, The Dead Sea – Jordan (abstract accepted).
5. **Shadia J Ikhmayies** and Riyad N Ahmad-Bitar, Thickness Dependence of the Bandgap Energy and Urbach Tail for CdS Thin Films Prepared by Vacuum Evaporation, *Eleventh World Renewable Energy Congress and Exhibition*, Abu Dhabi, United Arab Emirates 25–30 September (2010). (accepted).
6. **Shadia J Ikhmayies** and Riyad N Ahmad-Bitar, An Investigation of the Bandgap and Urbach Tail of Vacuum-Evaporated SnO_2Thin Films, *Eleventh World Renewable Energy Congress and Exhibition*, Abu Dhabi, United Arab Emirates 25–30 September (2010). (accepted).
7. **Shadia J Ikhmayies** and Riyad N Ahmad-Bitar, Temperature dependence of the PL-spectra of spray-deposited CdS:In thin films, Proceedings of the *Fifth International Conference on Thermal Engineering: Theory and Applications*, Marrakesh, Morocco May 10-14 (2010): 279-283.
8. **Shadia J Ikhmayies** and Riyad N Ahmad-Bitar, I-V Characterization of Selected Contacts for SnO_2:F Thin Films, *Fifth International Conference on Thermal Engineering: Theory and Applications*, Marrakesh, Morocco May 10-14 (2010): 264-268.
9. **Shadia J Ikhmayies** and Riyad N Ahmad-Bitar, Characterization of nanocrystalline CdS:In thin films prepared by the spray pyrolysis technique, *Collected Proceedings of the TMS 2010 139th Annual Meeting & Exhibition,* Seattle, Whashington, USA. February 14-18, (2010): 241-248.
10. **Shadia J Ikhmayies** and Riyad N Ahmad-Bitar, Using transmittance measurements to investigate the interdiffusion through the SnO_2/CdS and CdS/CdTe interfaces in SnO_2/CdS/CdTe solar cells, *Collected Proceedings of TMS 2010 139th Annual Meeting & Exhibition,* Seattle, Whashington, USA. February 14-18, (2010):109-117.
11. **Shadia J Ikhmayies** and Riyad N Ahmad-Bitar, A comparison between the electrical and optical properties of CdS:In thin films for two doping ratios, *Proceedings of the International Conference and Exhibition on Green Energy and Sustainability for Arid Regions and Mediterranean Countries (ICEGES 2009)*, Amman, Jordan, November 10-12 (2009).
12. **Shadia J Ikhmayies** and Riyad N Ahmad-Bitar, I-V characterization of selected contacts for CdS:In thin films, *Proceedings of the International Conference and Exhibition on Green Energy and Sustainability for Arid Regions and Mediterranean Countries (ICEGES 2009)*, Amman, Jordan, November 10-12 (2009).

13. **Shadia J Ikhmayies**, Naseem M. Abu El-Haija and Riyad N Ahmad-Bitar, Electrical and optical properties of spray-deposited ZnO:Al thin film, *Proceedings of the Global Conference on Renewables and Energy Efficiency for DEsert Regions*, Amman – Jordan, March 31st -April 2nd, (2009).
14. **Shadia J Ikhmayies** and Riyad N Ahmad-Bitar, Effect of the substrate temperature on the quality of spray-deposited CdS:In thin films, *Global Conference on Renewables and Energy Efficiency for DEsert Regions* , Amman – Jordan, March 31st -April 2nd, (2009).
15. **Shadia J Ikhmayies** and Riyad N Ahmad-Bitar, Characterization of the SnO_2:F/CdS:In Structures, *Collected Proceedings of TMS 2009 138th Annual Meeting & Exhibition*, San Francisco, California, USA, February 15-19, (2009): 399-407.
16. **Shadia J Ikhmayies** and Riyad N Ahmad-Bitar, Electrical, optical and structural properties of vacuum evaporated CdTe thin films, *Collected Proceedings of TMS 2009 138th Annual Meeting & Exhibition*, San Francisco, California, USA, February 15-19 (2009): 427-434.
17. **Shadia J. Ikhmayies**, Naseem M. Abu El-Haija and Riyad N. Ahmad-Bitar, Photovoltaic properties of undoped ZnO thin films prepared by the spray pyrolysis technique, *Proceedings of the Fourth International Conference on Thermal Engineering: Theory and Applications*, Abu Dhabi, UAE, January 12-14 (2009).
18. **Shadia J. Ikhmayies**, Naseem M. Abu El-Haija and Riyad N. Ahmad-Bitar, The influence of annealing in nitrogen atmosphere on the electrical, optical and structural properties of spray-deposited ZnO thin films, *Proceedings of the Fourth International Conference on Thermal Engineering: Theory and Applications*, Abu Dhabi, UAE, January 12-14 (2009).

In: Superconductivity and Condensed Matter ...
Editor: Collin J. Eckles

ISBN 978-1-61209-785-5

Chapter 13

BIOGRAPHICAL SKETCH – Superconductivity Research

NAME:	TITLE:	DATE OF BIRTH:
IVANSHIN, VLADIMIR A.	DR.	19.09.1957

EDUCATION

Institution and Location	Degree	Year Conferred	Scientific Field
Kazan State University, Russia	Diploma of the "engineer-radiophysicist"	1979	Radophysics and Electronics
Kazan State University, Russia	Ph.D	1989	Solid State Physics
Kazan Physical-Technical Institute, Russia	Habilit. (Dr. Sc.)	2008	Physics of Magnetism

CONTACT POINTS

MRS Laboratory, Faculty of Physics, Kazan State University, Kremlevskaya str., 18, 420008 Kazan, Tatarstan, Russia
e-mail: Vladimir.Ivanshin@ksu.ru
Phones:
Office: (007)(843)2337661
mobile: (007)(927)2429183
fax: (007)(843)2387201

RESEARCH AND PROFESSIONAL EXPERIENCE:

PROFESSIONAL APPOINTMENTS

Senior scientific researcher at the MRS Laboratory, Kazan State University, Russia, 2008 – present

Kazan State University, D. Sc. Student at the Department of Quantum Electronics and Radiospectroscopy, November 1, 2003 –October 30, 2006.

Scientific researcher at the MRS Laboratory, Kazan State University, 1989-2003; 2006-2007

Kazan State University, post graduate student at the Department of Quantum Electronics and Radiospectroscopy, 1985 - 1988;

University of Leipzig, Germany, Post-Doc. at the Institute of Physics, 1989

Kazan State University, MRS Laboratory, laboratory assistant, engineer, Junior scientific researcher, 1979 - 1993

HONORS RELEVANT TO RESEARCH:

1. DAAD-Stipendium at the University of Augsburg, Germany (Insitute of Experimental Physics IV, Prof. Dr. A. Loidl) October 2006-December 2006.
2. Scholarship of the Max Planck Society at the Max Planck Institute of Chemical Physics in Solids (Dresden, Germany) – February 2002 - February 2003 (German partner: Prof. Dr. F. Steglich)
3. DAAD-Stipendium at the Technische Hochschule Darmstadt, Germany (Insitute of Physical Chemistry, Prof. Dr. K.P. Dinse) – October 1996 – December 1996
4. Scholarship of the Federal Commission for Foreigners at the Swiss Federal Institute of Technology, Zurich, Switzerland (Institute of Quantum Electronics, Prof. Dr. P. Günter & Prof. Dr. J. Hulliger) – November 1991 – September 1992

PUBLICATIONS DURING LAST THREE YEARS

1. V.A. Ivanshin, I.N. Kurkin, A.M. Leushin, L.K. Aminov, "Crystal electric field excitations in the non-Fermi liquid compound $YbRh_2Si_2$", Journal of Superconductivity and Novel Magnetism.-2007.-V. 20, №2.-pp.131-133.
2. A.M. Leushin, V.A. Ivanshin, I.N. Kurkin, "Crystal field of Yb^{3+} tetragonal centers in the intermetallic compound $YbRh_2Si_2$", Physics of the Solid State .-2007.-V.49, N8.-pp. 1417 -1421.
3. V. A. Ivanshin, "ESR Bottleneck Effect in the Heavy-Fermion Metal YbRh2Si2", J. Magn. Magn. Mater.-2007.-V.316, pp. e393-e395.
4. A.M. Leushin, V. A. Ivanshin. Crystalline electric fields and the ground state of YbRh2Si2 and YbIr2Si2. Physica B, V.403,№5-9.-PP.1265-1267 (2008).
5. V.A. Ivanshin, A.A. Sukhanov, D.A. Sokolov, M.C. Aronson, S. Jia, S.L. Bud'ko, P.C. Canfield. Electron spin resonance of dense Yb-based heavy-fermion compounds: new experimental data, Journal of Alloys and Compounds, Vol. 480.-P. 126-127 (2008).

6. V. A. Ivanshin, I.N. Kurkin, E. V. Pomjakushina. Electron paramagnetic resonance of Ce^{3+} and Nd^{3+} impurity ions in $YBa_2Cu_3O_{6.13}$, Physics of the Solid State.-2009.-V.51, N2.-pp. 322 -326.
7. V. A. Ivanshin , T. O. Litvinova, A.A. Sukhanov, D.A. Sokolov, M.C. Aronson. Electron spin resonance in the Heusler alloy $YbRh_2Si_2$", JETP Lett. , V. 90, №2, pp. 116-119 (2009).
8. V. A. Ivanshin , T. O. Litvinova, and A.A. Sukhanov, Electronic hybridization effects in dense intermetallics measured by electron spin resonance, Journal of Physics: Conference series, in press

Other Contributions of Interest to Research: Projects, Grants etc.

1. SCOPES-Project of the Swiss National Science Foundation No. IB7420-110784 "Nanoscale properties of high-temperature superconducting cuprates and related compounds probed by magnetic resonance techniques" (Swiss partner: Prof. Dr. Hugo Keller, University of Zurich) – October 2005 – September 2008
2. SCOPES-Project of the Swiss National Science Foundation No. 7SUPJ062258 "Magnetic resonance studies on high-temperature superconducting cuprates and related compounds. Theory and experiment" (Swiss partner: Prof. Dr. Hugo Keller, University of Zurich) - August 2000 - July 2003
3. DFG-Project 436 RUS 113/566/0 „ESR-Experimente an Manganaten und theoretische Modelle" (German partner: Prof. Dr. A. Loidl, University of Augsburg, Germany) - October 1999 – March 2001
4. SEES/NIS Project of the Swiss National Science Foundation No. 7SUPJ048660 "Experimental study of high-temperature superconducting compounds by means of electron paramagnetic resonance and microwave absorption" (Swiss partner: Prof. Dr. Hugo Keller, University of Zurich) - December 1996 - May 1998

In: Superconductivity and Condensed Matter …
Editor: Collin J. Eckles
ISBN 978-1-61209-785-5

Chapter 14

BIOGRAPHICAL SKETCH – Superconductivity Research

NAME:	TITLE:	DATE OF BIRTH:
EKATERINA KALABUKHOVA	DSC	20.11.1947

EDUCATION

Institution and Location	Degree	Year Conferred	Scientific Field
National Technical University of Ukraine "Kiev Polytechnic Institute", Kiev, Ukraine	Engineer	1971	Electronics
V.E. Lashkaryov Institute of Semiconductor Physics National Academy of Sciences of Ukraine (NASU), Kiev, Ukraine	Candidate of science (PhD)	1979	Solid State Physics
V.E. Lashkaryov Institute of Semiconductor Physics NASU, Kiev, Ukraine	Doctor of Sciences	2005	Solid State Physics

CONTACT POINTS:

app. 102, h. 12, pr. Peremohy 37, Kiev, 03056, Ukraine,
tel: +380 44 236 09 71, fax: +380 44 489 17 04, e-mail: katia@i.kiev.ua

RESEARCH AND PROFESSIONAL EXPERIENCE

PROFESSIONAL APPOINTMENTS

Since 2005 - Leading Research Fellow at the V.E. Lashkaryov Institute of Semiconductor Physics NASU
1997-2004 - Head of the Wide Band Gap Semiconductor Material Sciences Department in the Special Design Technology Office of V.E. Lashkaryov Institute of Semiconductor Physics NASU
1988-2004 - Senior Research Fellow in V.E. Lashkaryov Institute of Semiconductor Physics NASU
1980-1988 - Reseach Fellow in V.E. Lashkaryov Institute of Semiconductor Physics NASU
1975-1980 - Engineer in V.E. Lashkaryov Institute of Semiconductor Physics NASU
1972-1975 - PhD Student in V.E. Lashkaryov Institute of Semiconductor Physics NASU

HONORS RELEVANT TO RESEARCH

PUBLICATIONS DURING LAST THREE YEARS

1. V.Yu. Ivanov, M. Godlewski, E.N. Kalabukhova, C.A. Dimitriadis, K. Zekentes, "Photoluminescence and optically detected magnetic resonance investigations of wurtzite phase 4H–SiC" Optical Materials, 2008, Vol. 30, pp. 746-748.
2. D.V. Savchenko, A. Pöppl, J. Hoentsch, E.N. Kalabukhova, Y. Bulois, A. Kassiba "Intrinsic defects in SiC nanoparticles as studied by pulsed electron paramagnetic resonance" Sol. St. Comm., 2008, Vol. 146, pp. 83–87.
3. A.V. Vasin, E.N. Kalabukhova, S.N. Lukin, D.V. Savchenko, V.S. Lysenko, A.N. Nazarov, A.V. Rusavsky, Y. Koshka "High-frequency electron paramagnetic resonance study of the as deposited and annealed carbonrich hydrogenated amorphous silicon-carbon films" Mater. Res. Soc. Symp. Proc., 2008, Vol. 1069, pp. 175-180.
4. I. Tkach, A. Baldansuren, E. Kalabukhova, S. Lukin, A. Sitnikov, A. Tsvir, M. Ischenko, Yu. Rosentzweig, E. Roduner «A Homebuilt ESE Spectrometer on the Basis of a High-Power Q-Band Microwave Bridge», Appl. Magn. Reson., 2008, Vol. 35, pp. 95-112.
5. D.V. Savchenko, B.D. Shanina, S.N. Lukin, E.N. Kalabukhova "Kinetics of the behavior of photosensitive impurities and defects in high-purity semi-insulating silicon carbide" Physics of the Solid State, 2009, Vol. 51, pp. 733–740.
6. D.V. Savchenko, A. Pöppl, E.N. Kalabukhova, S. Greulich-Weber, E. Rauls, W.G. Schmidt, U. Gerstmann "Spin-coupling in heavily nitrogen-doped 4H-SiC" Mat. Sci. Forum, 2009, Vols. 615-617, pp. 343-346.
7. D.V. Savchenko, E.N. Kalabukhova "EPR diagnostics of defect and impurity distribution homogeneity in semi-insulating 6H-SiC" Ukr. J. Phys. 2009, Vol. 54, pp. 605-610.

8. D.V. Savchenko, E.N. Kalabukhova, V.S. Kiselev, J. Hoentsch, A. Pöppl "Spin-Coupling and Hyperfine Interaction of the Nitrogen Donors in 6H-SiC', Phys. Status Solidi B, 2009, Vol. 246, pp. 1908-1914.
9. V.S. Kiselov, E.N. Kalabukhova, A.A. Sitnikov, P.M. Lytvyn, V.I. Poludin, V.O. Yukhymchuk, A.E. Belyaev «Effect of Si infiltration method on the properties of biomorphous SiC», SPQEO, 2009. Vol. 12, pp. 68-71
10. D.V. Savchenko, E.N. Kalabukhova, S.N. Lukin, E.N. Mokhov, J. Hoentsch, A. Pöppl "EPR, ESE and pulsed ENDOR study of the nitrogen donor pairs on quasi-cubic lattice sites in 6H SiC" Physica B, 2009, Vol. 404, pp. 4735-4738.
11. V.S. Kiselov, E.N. Kalabukhova, A.A. Sitnikov, P.M. Litvin, V.I. Poludin, V.A. Yukhymchyk, A.E. Belyaev "Effect of Si infiltration method on the biomorphous SiC microstructure properties", Nanosystems, nanomaterials, nanotechologies, 2009, Vol. 7, pp. 237-243.
12. D. V. Savchenko, A. Pöppl, E. N. Kalabukhova, E. F. Venger, M. P. Gadzira, G. G. Gnesin "Intrinsic Defects in Nonstoichiometric b-SiC Nanoparticles Studied by Pulsed Magnetic Resonance Methods", SPQEO, 2010, Vol. 13, pp. 43-50.
13. E.N. Kalabukhova, S.N. Lukin, D.V. Savchenko, B.D. Shanina, A.V. Vasin, V.S. Lysenko, A.N. Nazarov, A.V. Rusavsky, J. Hoentsch, Y. Koshka "The EPR study of the carbon and silicon related defects in carbon-rich hydrogenated amorphous silicon-carbon films", Phys. Rev. B, 2010, Vol. 88, pp. 155319-1-155319-9.

OTHER CONTRIBUTIONS OF INTEREST TO RESEARCH: PROJECTS, GRANTS ETC.

1992-2000 - Visiting Scientist in Laboratory of Radiospectroscopy, University of Paderborn (Germany) - every year, total time of about 3 years

1997, 1999, 2004, 2005 - USA visits on program "Window of Science"

2004-2006 – INTAS project # 03-51-3747 "High Field multifrequency EPR and ODEPR study of the deep intrinsic defects in semi-insulating 6H and 4H SiC material".

2005-2006 – STCU project #P236 "Elaboration of a pulse microwave bridge for a Q-band Electron Spin-Echo Spectrometer".

2007-2009 – CRDF project #UKE 2-28-56 "Research and Development of Amorphous Si_XC_{1-X}:H and Rare-Earth Doped Amorphous Si_XC_{1-X}:H films on Crystalline Si for Light-Emitting Devices"

2009-2012 – SCOPES project IZ73Z0_127945/1 "Spin interference processes under magnetic resonance of single point defects embedded in quantum wells, wires and rings"

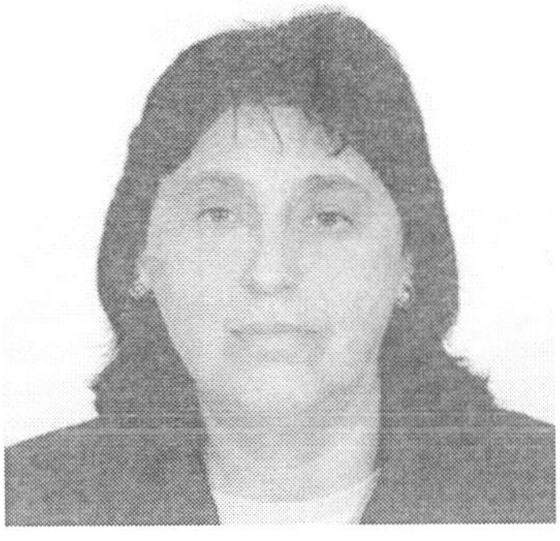

In: Superconductivity and Condensed Matter ...
Editor: Collin J. Eckles
ISBN 978-1-61209-785-5

Chapter 15

BIOGRAPHICAL SKETCH – Superconductivity Research

NAME:	TITLE:	DATE OF BIRTH:
CORNELIUS KRELLNER	DR.	AUGUST 29TH, 1978

EDUCATION

Institution and Location	Degree	Year Conferred	Scientific Field
Kreuzschule Dresden, Germany	Abitur (baccalaureate)	1997	
TU Dresden, Germany	Vordiplom	2002	Physics
ETH Zurich, Switzerland and TU Dresden, Germany	Dipl. Phys.	2004	Physics
TU Dresden	Dr. rer. Nat.	2009	Condensed matter physics, Material science

CONTACT POINTS

Max-Planck-Institute for Chemical Physics of Solids, Nöthnitzer Str. 40, D-01087 Dresden, Germany

http://www.cpfs.mpg.de/~krellner/

RESEARCH AND PROFESSIONAL EXPERIENCE

Since 2009: Postdoc at the Max-Planck-Institute for Chemical Physics of Solids, Dresden Germany

PROFESSIONAL APPOINTMENTS

Referee of scientific journals, e.g. Physical Review Letters and Physical Review B

HONORS RELEVANT TO RESEARCH

Otto Hahn Medal 2010 of the Max Planck Society.
ThyssenKrupp Electical Steel Dissertationspreis 2010 of the German Physical Society.
Stipendiary of the DAAD (2002/2003)

PUBLICATIONS DURING LAST THREE YEARS

http://www.cpfs.mpg.de/~krellner/)

Author and Co-author of more than 50 articles in reviewed journals: 1 in Science, 1 in Nature Physics, and 9 in Physical Review Letters

h-Index: 14, more than 600 citations

PhD-thesis "Ferromagnetische Korrelationen in Kondo-Gittern: YbT2Si2 und CeTPO (T=Übergangsmetall)" Mensch und Buch Verlag 2009, Berlin (ISBN: 3866646526)

1. S. Friedemann, N. Oeschler, S. Wirth, **C. Krellner**, C. Geibel, F. Steglich, S. Paschen, S. Kirchner, and Q. Si, **PNAS** in press, (2010).
2. M. Holder, A. Jesche, P. Lombardo, R. Hayn, D. V. Vyalikh, S. Danzenbacher, K. Kummer, **C. Krellner**, C. Geibel, Yu. Kucherenko, T. K. Kim, R. Follath, S. L. Molodtsov, and C. Laubschat, *CeFePO: f-d hybridization and quenching of superconductivity,* **Phys. Rev. Lett. 104,** 096402 (2010).
3. S. Hartmann, N. Oeschler, **C. Krellner**, C. Geibel, S. Paschen, and F. Steglich, *Thermopower evidence for an abrupt Fermi surface change at the quantum critical point of* $YbRh_2Si_2$, **Phys. Rev. Lett. 104,** 096401 (2010).
4. D.V. Vyalikh, S. Danzenbächer, Yu. Kucherenko, **C. Krellner**, C. Geibel, C. Laubschat, M. Shi, L. Patthey, R. Follath, and S. L. Molodtsov, *Tuning the Hybridization at the Surface of a Heavy-Fermion System*, **Phys. Rev. Lett. 103**, 137601 (2009).
5. S. Friedemann, T. Westerkamp, M. Brando, N. Oeschler, S. Wirth, P. Gegenwart, **C. Krellner**, C. Geibel, and F. Steglich, *Detaching the antiferromagnetic quantum critical point from the Fermi-surface reconstruction in* $YbRh_2Si_2$, Nature Physics 5, 465 (2009).
6. **C. Krellner**, S. Hartmann, A. Pikul, N. Oeschler, J. G. Donath, C. Geibel, F. Steglich, and J. Wosnitza, *Violation of Critical Universality at the Antiferromagnetic Phase Transition of* $YbRh_2Si_2$, **Phys. Rev. Lett. 102**, 196402 (2009).
7. U. Schaufuß, V. Kataev, A. A. Zvyagin, B. Büchner, J. Sichelschmidt, J. Wykhoff, **C. Krellner**, C. Geibel, and F. Steglich, *Evolution of the Kondo State of* $YbRh_2Si_2$ *Probed by High-Field ESR*, **Phys. Rev. Lett. 102**, 076405 (2009).

8. S. Danzenbächer, D. V. Vyalikh, Yu. Kucherenko, A. Kade, C. Laubschat, N. Caroca-Canales, **C. Krellner**, C. Geibel, A. V. Fedorov, D. S. Dessau, R. Follath, W. Eberhardt, and S. L. Molodtsov; *Hybridization Phenomena in Nearly-Half-Filled f-Shell Electron Systems: Photoemission Study of $EuNi_2P_2$*, **Phys. Rev. Lett. 102**, 026403 (2009). **On the Cover of Vol. 102 Iss.2.**
9. A. Jesche, N. Caroca-Canales, H. Rosner, H. Borrmann, A. Ormeci, D. Kasinathan, H. H. Klauss, H. Luetkens, R. Khasanov, A. Amato, A. Hoser, K. Kaneko, **C. Krellner**, and C. Geibel, *Strong coupling between magnetic and structural order parameters in $SrFe_2As_2$*, **Phys. Rev. B 78**, 180504(R) (2008)
10. E. M. Brüning, **C. Krellner**, M. Baenitz, A. Jesche, F. Steglich, and C. Geibel, *CeFePO: A Heavy Fermion Metal with Ferromagnetic Correlations*, **Phys. Rev. Lett. 101**, 117206 (2008)
11. **C. Krellner,** N. Caroca-Canales, A. Jesche, H. Rosner, A. Ormeci, and C. Geibel, *Magnetic and structural transitions in layered iron arsenide systems: AFe_2As_2 versus RFeAsO*, **Phys. Rev. B 78**, 100504(R) (2008)
12. **C. Krellner**, T. Förster, H. Jeevan, C. Geibel, and J. Sichelschmidt, *Relevance of ferromagnetic correlations for the Electron Spin Resonance in Kondo lattice systems*, **Phys. Rev. Lett. 100**, 066401 (2008)
13. D.V. Vyalikh, S. Danzenbächer, A.N. Yaresko, M. Holder, Yu. Kucherenko, C. Laubschat, **C. Krellner**, Z. Hossain, C. Geibel, M. Shi, L. Patthey, and S.L. Molodtsov; *Photoemission Insight into Heavy-Fermion Behavior in $YbRh_2Si_2$*, **Phys. Rev. Lett. 100**, 056402 (2008)
14. **C. Krellner**, S. Haas, C. Goldmann, K. P. Pernstich, D. J. Gundlach, and B. Batlogg, *Density of bulk trap states in organic semiconductor crystals: discrete levels induced by oxygen in rubrene*, **Phys. Rev. B 75**, 245115 (2007)
15. P. Gegenwart, T. Westerkamp, **C. Krellner**, Y. Tokiwa, S. Paschen, C. Geibel, F. Steglich, E. Abrahams, and Q. Si; *Multiple energy scales at a quantum critical point*, **Science 315**, 969 (2007)

OTHER CONTRIBUTIONS OF INTEREST TO RESEARCH: PROJECTS, GRANTS ETC.

THIRD-PARTY FUNDING

- **1 PhD student and consumables** within the priority program (3+3 years) "Hochtemperatursupraleitung in Eisenpniktiden" (SPP 1458) of the DFG (Deutsche Forschungsgemeinschaft) started in 07/2010

In: Superconductivity and Condensed Matter …
Editor: Collin J. Eckles
ISBN 978-1-61209-785-5

Chapter 16

BIOGRAPHICAL SKETCH – Superconductivity Research

NAME:	TITLE:	DATE OF BIRTH:
HUAN-CHIU KU (H. C. KU)	DEAN, COLLEGE OF SCIENCE	FEBRUARY 11, 1949, TAIPEI, TAIWAN, REPUBLIC OF CHINA

CONTACT POINTS

Dean, College of Science
National Tsing Hua University
Hsinchu, Taiwan 30013, Republic of China
Tel : 886-3-5719039 (CoS), 5715928 (Lab)
Fax: 886-3-5723052
e-mail: hcku@phys.nthu.edu.tw

EDUCATION

Ph.D. in Physics	University of California, San Diego (UCSD), USA (1975-1980) (Advisor: Professor B.T. Matthias)
M.S. in Physics	Ohio University, USA (1974-1975)
B.S. in Physics	National Cheng Kung University (1967-1971)

PROFESSIONAL EXPERIENCES

2007-	Dean, College of Science, National Tsing Hua University
2006-	Tsing Hua Chair Professor
2003-	Tsing Hua Chair Professor of Natural Sciences

1995-98	Director, Division of Natural Sciences and Mathematics, National Science Council
1990-93	Chairman and Director, Department and Institute of Physics, National Tsing Hua University
1986-87	Visiting Professor, Iowa State University, USA
1984-	Professor, Department of Physics, National Tsing Hua University
1980-84	Associate Professor, National Tsing Hua University
1980-81	Postgraduate Research Physicist, Institute of Pure and Applied Physics, University of California, San Diego, USA

HONORS AND AWARDS

2008	Honorary Visiting Professor, University of Science and Technology of China, PROC
2006-	Tsing Hua Chair Professor
2003-	Tsing Hua Chair Professor of Natural Sciences
2004	NSC Outstanding Research Fellow Award
2003	Special Contribution Award, Physical Society of Republic of China
2000-05	Outstanding Scholarship Award
1998-04	NSC Research Fellow
1997-	Fellow, Physical Society of Republic of China
1996	Honorary Professor, Henan Normal University, PROC
1995	Cinese Education & Culture Foundation Award
1992	Sun Yat-Sen Academic Prize
1991-92	NSC Outstanding Research Prize
1989-90	NSC Outstanding Research Prize
1987-88	National Science Council Outstanding Research Prize

RESEARCH AREA

Experimental Condensed Matter Physics, Low Temperature Physics, Superconductivity, Magnetism, Strongly Correlated Electron Systems, Material Sciences, and Nanomagnetism

RESEARCH PAPERS

250+ Referred Journal Papers (1979-2010) on Superconductivity, Magnetism and Low Temperature Physics

Including Iron-based Superconductors, Magnetic Superconductors, Heavy Fermion Superconductors, Valence Fluctuation Systems, New Magnetic Materials, New Superconducting Materials, High-T_c Cuprate Superconductors, CMR Materials, Metal-insulator Transition, Strongly-Correlated Electron Systems.

Research-Related Activities

2003-05	Director, Physics Research Promotion Center, National Science Council
1995-98	Director, Division of Natural Sciences and Mathematics, National Science Council
1990-93	Director, Physics Research Promotion Center, National Science Council
1989-90	Physics Project Coordinator, National Science Council

In: Superconductivity and Condensed Matter ...
Editor: Collin J. Eckles
ISBN 978-1-61209-785-5

Chapter 17

BIOGRAPHICAL SKETCH – Superconductivity Research

NAME:	TITLE:	DATE OF BIRTH:
FRANK STEGLICH	PROFESSOR DR. RER. NAT.	14/03/1941

EDUCATION

Institution and Location	Degree	Year Conferred	Scientific Field
Städtisches Jungengymnasium Bottrop/W	Abitur	1960	
University of Goettingen	Diploma	1966	Physics
U. Goettingen	Dr. rer. nat.	1969	Physics
U. Cologne	Habilitation	1976	Physics

RESEARCH AND PROFESSIONAL EXPERIENCE

PROFESSIONAL APPOINTMENTS

Honorary Professor, TU Dresden, since 1999
Director, Max-Planck Institute for Chemical Physics of Solids, Dresden, since 1996
Professor of Physics (C4), TU Darmstadt, 1980-98
Professor of Physics (C3), TH Darmstadt, 1978-80
Lecturer, U. Cologne, 1976-78
Research Fellow, U. Cologne, 1969-76
Research Fellow, U. Göttingen, 1966-69

HONORS RELEVANT TO SUPERCONDUCTIVITY RESEARCH

Leibniz Prize (DFG), 1986
Hewlett-Packard Europhysics Prize, 1989
AvH Award, 1989
APS Intern. Prize for New Materials, 1990
IUPAP Magnetism Award, 2000
Stern-Gerlach Medal (DPG), 2004
Bernd T. Matthias Prize for Superconducting Materials, 2006

PUBLICATIONS DURING LAST THREE YEARS

1. Influence Of Antiferromagnetic Fluctuations On The Fulde-Ferrell-Larkin- Ovchinnikov State In $CeCoIn_5$ (invited paper)
   M. Nicklas, C.F. Miclea, J.L. Sarrao, J.D. Thompson, G. Sparn and F. Steglich
   J. Low Temp. Phys. 146, 669 (2007).
2. Spatial Separation Of Antiferromagnetism And Superconductivity In $CeCu_2Si_2$
   E. Faulhaber, O. Stockert, K. Schmalzl, H.S. Jeevan, M. Deppe, G. Geibel, F. Steglich and M. Loewenhaupt
   J. Magn. Magn. Mat. 310, 295 (2007).
3. Low-Temperature Muon Spin Relaxation Measurements ON $CeRu_4Sn_6$
   A.M. Strydom, A.D. Hillier, D.T. Adroja, S. Paschen and F. Steglich
   J. Magn. Magn. Mat. 310, 377 (2007).
4. ^{119}Sn NMR On The Correlated Semi-Metal $CeRu_4Sn_6$
   E.M. Brüning, M. Baenitz, A.A. Gippius, A.M. Strydom, F. Steglich and R.E. Walstedt
   J. Magn. Magn. Mat. 310, 393 (2007).
5. Far-Infrared Optical Investigation Of The Heavy-Fermion Semiconductor U_2Ru_2Sn
   J. Sichelschmidt, V. Voevodin, J.A. Mydosh and F. Steglich
   J. Magn. Magn. Mat. 310, 434 (2007).
6. Structure And Transport Properties Of New Rare-Earth Gallium Telluride $YbGa_6T_{10}$
   A. Bentien, S. Budnyk, Yu. Prots, Yu. Grin and F. Steglich
   J. Alloys Comp. 442, 345 (2007).
7. Quantum Critical Behaviour In $Ce_3Pd_{20}Si_6$?
   S. Paschen, M. Müller, J. Custers, M. Kriegisch, A. Prokofiev, G. Hilscher, W. Steiner, A. Pikul, F. Steglich, A. Strydom
   J. Magn. Magn. Mat. 316, 90 (2007).
8. Metallic State In Cubic FeGe Beyond Its Quantum Phase Transition
   P. Pedrazzini, H. Wilhelm, D. Jaccard, T. Jarlborg, M. Schmidt, M. Hanfland, L. Akselrud, H.Q. Yuan, U. Schwarz, Yu. Grin and F. Steglich
   Phys. Rev. Lett. 98, 047204 (2007).
9. Probing The Quantum Critical Behavior Of $CeCoIn_5$ Via Hall Effect Measurements

S. Singh, C. Capan, M. Nicklas, M. Rams, A. Gladun, H. Lee, J.F. DiTusa, Z. Fisk, F. Steglich and S. Wirth
Phys. Rev. Lett. 98, 057001 (2007).

10. Superconductivity And Magnetism: From Antagonism To Mutual Interplay (invited paper)
F. Steglich
Physica C 460 - 462, 7 (2007).
11. Unconventional Superconductivity And Quantum Criticality In Pressurized $CeCu_2(Si_{1-x}Ge_x)_2$ (invited paper)
H.Q. Yuan and F. Steglich
Physica C 460 - 462, 141 (2007).
12. Pressure Effect On Antiferromagnetism In $CeRhIn_{5-x}Sn_x$ Studied By Thermal Expansion
J.G. Donath, P. Gegenwart, F. Steglich, E.D. Bauer and J.L. Sarrao
Physica C 460 - 462, 661 (2007).
13. Crossover From Landau Fermi Liquid To Non-Fermi Liquid Behavior: Indications From Hall Measurements On $CeCoIn_5$
C. Capan, S. Singh, S. Nair, M. Nicklas, H. Lee, J.F. DiTusa, Z. Fisk, S. Wirth and F. Steglich
Physica C 460 - 462, 678 (2007).
14. Electron Spin Resonance In $YbRh_2Si_2$: The Role Of The Residual Linewidth
J. Wykhoff, J. Sichelschmidt, J. Ferstl, C. Krellner, C. Geibel, F. Steglich, I. Fazlishanov and H.-A. Krug von Nidda
Physica C 460 - 462, 686 (2007).
15. Compact Hydrostatic Pressure Cell For Precise Resistivity Measurements Up To 10 GPa Using PPMS
T. Nakanishi, M. Nicklas, G. Sparn and F. Steglich
J. Phys. Soc. Jpn. 76 (Suppl. A), 223 (2007).
16. Electron Spin Resonance Of $YbIr_2Si_2$ Below The Kondo Temperature
J. Sichelschmidt, J. Wykhoff, H.-A. Krug von Nidda, I.I. Fazlishanov, Z. Hossain, C. Krellner, C. Geibel and F. Steglich
J. Phys.: Condens. Matter 19, 016211 (2007).
17. Short-Range Magnetic Ordering In $URh_{0.7}Ru_{0.3}Ge$
V.H. Tran, W. Miiller, M. Baenitz and F. Steglich
J. Phys.: Condens. Matter 19, 096201 (2007).
18. Spin Dynamics Of $YbRh_2Si_2$ Observed By Electron Spin Resonance
J. Sichelschmidt, J. Wykhoff, H.-A. Krug von Nidda, J. Ferstl, C. Geibel and F. Steglich
J. Phys.: Condens. Matter 19, 116204 (2007).
19. Low-Temperature Study Of The Strongly Correlated Compound $Ce_3Rh_4Sn_{13}$
U. Köhler, A.P. Pikul, N. Oeschler, T. Westerkamp, A.M. Strydom and F. Steglich
J. Phys.: Condens. Matter 19, 386207 (2007).
20. Antiferromagnetic Fluctuations And The Fulde-Ferrell-Larkin- Ovchinnikov State in $CeCoIn_5$ (invited paper)
M. Nicklas, C.F. Miclea, J.L. Sarrao, J.D. Thompson, G. Sparn and F. Steglich

J. Phys. Soc. Jpn. 76 (Suppl. A), 128 (2007).

21. Resistivity Anomaly In Structurally Disordered Pbfcl-Type Arsenide Selenides (invited paper)
T. Cichorek, D. Gnida, R. Niewa, A. Schlechte, M. Schmidt, Yu. Prots, R. Ramlau, Z. Henkie, R. Kniep and F. Steglich
J. Low Temp. Phys. 147, 309 (2007).
22. Multiple Energy Scales At A Quantum Critical Point
P. Gegenwart, T. Westerkamp, C. Krellner, Y. Tokiwa, S. Paschen, C. Geibel, F. Steglich, E. Abrahams and Q. Si
Science (Reports) 315, 969 (2007).
23. Scanning Tunneling Spectroscopy On $Pr_{0.68}Pb_{0.32}MnO_3$ Single Crystals
S. Rößler, S. Ernst, B. Padmanabhan, Suja Elizabeth, H.L. Bhat, S. Wirth and F. Steglich
IEEE Trans. Magn. 43 (2007).
24. Magnetism And Quantum Criticality In Heavy-Fermion Compounds: Interplay With Superconductivity (invited paper)
O. Stockert, M. Nicklas, P. Thalmeier, P. Gegenwart and F. Steglich
Handbook on Magnetism and Advanced Magnetic Materials (Wiley, New York, 2007), p. 2461.
25. Colossal Seebeck Coefficient In Strongly Correlated Semiconductor $FeSb_2$
A. Bentien, S. Johnsen, G.K.H. Madsen, B.B. Iversen and F. Steglich
Europhys. Lett. 80, 17008 (2007).
ERRATUM
Europhys. Lett. 80, 39901 (2007).
26. Optical Evidence For Heavy Charge Carriers In FeGe
V. Guritanu, D. van der Marel, J. Teyssier, T. Jarlborg, H. Wilhelm, M. Schmidt and F. Steglich
Phys. Rev. B 75, 155114 (2007).
27. Nanometer-Scale Phase Separation In A Colossal Magnetoresistive Manganite
S. Rößler, S. Ernst, B. Padmanabhan, S. Elizabeth, H.L. Bhat, F. Steglich and S. Wirth
Europhys. Lett. 83, 17009 (2008).
28. Magnetic Structure Of Cd-DOPED $CeCoIn_5$
M. Nicklas, O. Stockert, T. Park, K. Habicht, K. Kiefer, L.D. Pham, J. D. Thompson, Z. Fisk and F. Steglich
Phys. Rev. B 76, 052401 (2007).
29. Chemical Physics Of Solids
Y. Grin, R. Kniep and F. Steglich
Sci. Technol. Adv. Mat. 8, 339 (2007).
30. Thermo-Chemical Properties And Electrical Resistivity Of Zr-Based Arsenide Chalcogenides (invited paper)
A. Schlechte, R. Niewa, M. Schmidt, G. Auffermann, Yu. Prots, W. Schnelle, D. Gnida, T. Cichorek, F. Steglich and R. Kniep
Sci. Technol. Adv. Mat. 8, 341 (2007).
31. Tip Preparation For Usage In An Ultra-Low Temperature UHV Scanning Tunneling Microscope (invited paper)
S. Ernst, S. Wirth, M. Rams, V. Dolocan and F. Steglich

Sci. Technol. Adv. Mat. 8, 347 (2007).
32. Low-Energy Spin Fluctuations In The Non-Fermi-Liquid Compound $YbRh_2Si_2$ (invited paper) Stockert, M.M. Koza, J. Ferstl, C. Geibel and F. Steglich Sci. Technol. Adv. Mat. 8, 371 (2007).
33. On The Local And Itinerant Properties Of The Esr In $YbRh_2Si_2$ (invited paper)
J. Wykhoff, J. Sichelschmidt, G. Lapertot, G. Knebel, J. Flouquet, I.I. Fazlishanov, H.-A. Krug von Nidda, C. Krellner, C. Geibel and F. Steglich
Sci. Technol. Adv. Mat. 8, 389 (2007).
34. Field-Induced Magnetic Phase Transition As A Magnon Bose Einstein Condensation (invited paper)
T. Radu, Y. Tokiwa, R. Coldea, P. Gegenwart, Z. Tylczynski and F. Steglich
Sci. Technol. Adv. Mat. 8, 406 (2007).
35. Systematic Study Of The Grüneisen Ratio Near Quantum Critical Points (invited paper)
R. Küchler, P. Gegenwart, C. Geibel and F. Steglich
Sci. Technol. Adv. Mat. 8, 428 (2007).
36. $R_3Ir_4Ge_{13}$ (R = Yb, Lu): Thermal And Magnetic Properties
A.M. Strydom, N. Oeschler and F. Steglich
Physica B 403, 746 (2008).
37. Optical Properties Of $YbRh_2Si_2$ And $YbIr_2Si_2$: A Comparison
J. Sichelschmidt, S. Kimura, C. Krellner, C. Geibel and F. Steglich
Physica B 403, 775 (2008).
38. Hall Effect Measurements In The Heavy Fermion System $CeIrIn_5$
S. Nair, S. Wirth, M. Nicklas, A. Gladun, F. Steglich, J.L. Sarrao and J.D. Thompson
Physica B 403, 837 (2008).
39. Pressure Dependence Of The Neel And The Superconducting Transition Temperature Of $CeCo(In_{0.9}Cd_{0.1})_5$ Studied By Thermal Expansion
J.G. Donath, P. Gegenwart, M. Nicklas, F. Steglich, L.D. Pham and Z. Fisk
Physica B 403, 839 (2008).
40. Heat Capacity Studies Of Single-Crystalline $CePt_4In$
A.P. Pikul, D. Kaczorowski, Z. Bukowski and F. Steglich
Physica B 403, 842 (2008).
41. Experimental Evidence For Unconventional BCS States In Heavy- Fermion Metals (invited paper)
F. Steglich, C. Geibel, F.M. Grosche, M. Loewenhaupt, O. Stockert, S. Wirth and H.Q. Yuan
Physica B 403, 968 (2008).
42. Magnetism And Superconductivity In The Heavy-Fermion Compound $CeCu_2Si_2$ Studied By Neutron Scattering (invited paper)
O. Stockert, J. Arndt, A. Schneidewind, H. Schneider, H.S. Jeevan, C. Geibel, F. Steglich and M. Loewenhaupt
Physica B 403, 973 (2008).
43. Unconventional Quantum Criticality In $YbRh_2Si_2$ (invited paper)

P. Gegenwart, T. Westerkamp, C. Krellner, M. Brando, Y. Tokiwa, C. Geibel and F. Steglich
Physica B 403, 1184 (2008).

44. Low-Temperature Magnetic Susceptibility Of $Yb(Rh_{1-x}M_x)_2Si_2$ (M = Ir, Co) Single Crystals
T. Westerkamp, P. Gegenwart, C. Krellner, C. Geibel and F. Steglich
Physica B 403, 1236 (2008).

45. Band-Structure And Anomalous Contributions To The Hall Effect In $YbRh_2Si_2$
S. Friedemann, N. Oeschler, C. Krellner, C. Geibel, S. Wirth, F. Steglich, S. Paschen, S. MaQuilon and Z. Fisk
Physica B 403, 1251 (2008).

46. Low-Temperature Specific Heat Of $YbRh_2Si_2$
N. Oeschler, S. Hartmann, A.P. Pikul, C. Krellner, C. Geibel and F. Steglich
Physica B 403, 1254 (2008).

47. New Hints On The Origin Of Quantum Criticality In $CeCoIn_5$: A Hall Effect Study
C. Capan, S. Singh, S. Wirth, M. Nicklas, H. Lee, Z. Fisk, J. DiTusa and F. Steglich
Physica B 403, 1290 (2008).

48. Hall Effect Of $YbIr_2Si_2$
M. Kriegisch, J. Custers, C. Geibel, G. Hilscher, C. Krellner, M. Müller, F. Steglich and S. Paschen
Physica B 403, 1295 (2008).

49. Dimensional Crossover Of Quantum Critical Behavior In $CeCoIn_5$
J.G. Donath, F. Steglich, E.D. Bauer, J.L. Sarrao and P. Gegenwart
Phys. Rev. Lett. 100, 136401 (2008).

50. NMR Investigations Of U_2Ru_2Sn: A Possible Kondo Insulator
A.K. Rajarajan, A. Rabis, M. Baenitz, A.A. Gippius, E.N. Morozowa, J.A. Mydosh and F. Steglich
Phys. Rev. B 76, 024424 (2007).

51. Colossal Seebeck Coefficient In Semiconducting $FeSb_2$ (invited paper)
A. Bentien, S. Johnson, G.K.H. Madsen, B.B. Iversen and F. Steglich
Europhys. News 38, 19 (2007).

52. Spin Fluctuations In The Ferromagnets (Na, K)$FeSb_{12}$ And The Non-Ordered Homologues (Ba, Ca)Fe_4Sb_{12} -Charge And Thermal Transport
S. Narayani, A. Leithe-Jasper, M. Baenitz, Y. Grin and F. Steglich
Proc. 52nd DAE Solid State Physics Symposium (Univ. of Mysore, Mysore, India, 2007) p. 1093.

53. Density Of States And Spatially Inhomogeneous Conductance Near The Metal-Insulator Transition In $Pr_{0.68}Pb_{0.32}MnO_3$ Single Crystals
S. Wirth, S. Rößler, S. Ernst, B. Padmanabhan, H.L. Bhat, S. Elizabeth and F. Steglich
J. Phys.: Condens. Matter 20, 434231 (2008).

54. Precursor State To Unconventional Superconductivity In $CeIrIn_5$
S. Nair, S. Wirth, M. Nicklas, J.L. Sarrao, J.D. Thompson, Z. Fisk and F. Steglich

Phys. Rev. Lett. 100, 137003 (2008).

55. Energy Scales Of $Lu_{1-x}Yb_xRh_2Si_2$ By Means Of Thermopower Investigations
    U. Köhler, N. Oeschler, F. Steglich, S. MaQuilon and Z. Fisk
    Phys. Rev. B 77, 104412 (2008).
56. Quantum Criticality In Heavy-Fermion Metals (invited paper)
    P. Gegenwart, Q. Si and F. Steglich
    Nature Phys. 4, 186 (2008).
57. CeFePO: A Heavy-Fermion Metal With Ferromagnetic Correlations
    E. M. Brüning, C. Krellner, M. Baenitz, A. Jesche, F. Steglich and C. Geibel
    Phys. Rev. Lett. 101, 117206 (2008).
58. Evolution Of The Kondo State Of $YbRh_2Si_2$ Probed By High Field ESR
    U. Schaufuß, V. Kataev, A.A. Zvyagin, B. Büchner, J. Sichelschmidt, J. Wykhoff, C. Krellner, C. Geibel and F. Steglich
    Phys. Rev. Lett. 102, 076405 (2009).
59. High Pressure NQR Measurement In $CeCu_2Si_2$ Up To Sudden Disappearance Of Superconductivity K. Fujiwara, Y. Hata, K. Kobayashi, K. Miyoshi, J. Takeuchi, Y. Shimaoka, H. Kotegawa, T.C. Kobayashi, C. Geibel and F. Steglich
    J. Phys. Soc. Jpn. 77, 123711 (2008).
60. Narrow-Gap-Related Thermoelectric Enhancement In $FeSb_2$
    P. Sun, N. Oeschler, F. Steglich, S. Johnsen and B.B. Iversen
    Applied Energy (in press).
61. Thermal Conductivity Reduction Of Colossal Power Factor Compound $FeSb_2$ Through Compensated Element Substitution
    S. Johnsen, G.K.H. Madsen, B.B. Iversen, A. Bentien, P. Sun, N. Oeschler and F. Steglich
    Applied Energy (in press).
62. The Effect Of Pressure And Ir Substitution In $YbRh_2Si_2$
    M. Macovei, M. Nicklas, C. Krellner, C. Geibel and F. Steglich
    J. Phys.: Condens. Matter 20, 505205 (2008).
63. Thermoelectric Properties Of The Narrow-Gap Semiconductors $FeSb_2$ AND $RuSb_2$: A Comparative Study
    P. Sun, N. Oeschler, S. Johnsen, B.B. Iversen and F. Steglich
    J. Phys. Conference Series 150, 012049 (2009).
64. Do Antiferromagnetism And Superconductivity Coexist In 2 % And 10 % Ge Doped $CeCu_2Si_2$?
    J. Arndt, O. Stockert, R. Borth, E. Faulhaber, K. Schmalz, A. Schneidewind, H.S. Jeevan, C. Geibel, M. Loewenhaupt and F. Steglich
    J. Phys. Conference Series 150, 042008 (2009).
65. Magnetic Resonance On Correlated Semimetals: The Case Of U_2Ru_2Sn, $CeRu_4Sn_6$ AND $FeSb_2$
    A.A. Gippius, M. Baenitz, A.K. Rajarajan, E.M. Brüning, K. Okhotnikov, R. Walstedt, A. Strydom, J. Mydosh and F. Steglich
    J. Phys. Conference Series 150, 042040 (2009).

66. Low-Temperature Thermopower Study Of $YbRh_2Si_2$
S. Hartmann, N. Oeschler, C. Krellner, C. Geibel and F. Steglich
J. Phys. Conference Series 150, 042049 (2009).
67. High-Field Esr Study Of The Kondo Lattice System $YbRh_2Si_2$
V. Kataev, U. Schaufuß, B. Büchner, A.A. Zvyagin, J. Sichelschmidt, J. Wykhoff, C. Krellner, C. Geibel and F. Steglich
J. Phys. Conference Series 150, 042085 (2009).
68. Field Dependence Of The Thermopower Of CeNiSn
U. Köhler, P. Sun, N. Oeschler, T. Takabatake, S. Paschen and F. Steglich
J. Phys. Conference Series 150, 042096 (2009).
69. Hall Effect And Magnetoresistance In The Heavy Fermion Superconductor $CeCo(In_{1-x}Cd_x)_5$
S. Nair, S. Wirth, M. Nicklas, A.D. Bianchi, Z. Fisk and F. Steglich
J. Phys. Conference Series 150, 042133 (2009).
70. Influence Of Microstructure On Local Conductivities In $La_{0.7}Ce_{0.3}MnO_3$ Thin Film
S. Rößler, J. Jesudasan, K. Bajaj, P. Raychaudhuri, F. Steglich and S. Wirth
J. Phys. Conference Series 150, 042164 (2009).
71. Magnetization Measurements On $YbRh_2Si_2$ At Very Low Temperatures
E. Schuberth, M. Tippmann, M. Kath, C. Krellner, C. Geibel, T. Westerkamp, C. Klingner and F. Steglich
J. Phys. Conference Series 150, 042178 (2009).
72. Quantum Criticality In Layered $CeRhIn_{5-x}Sn_x$ Compared With Cubic $CeIn_{3-x}Sn_x$
J. G. Donath, F. Steglich, E.D. Bauer, F. Ronning, J.L. Sarrao and P. Gegenwart
Europhys. Lett. 87, 57011 (2009).
73. Precursor State To Superconductivity In $CeIrIn_5$: Unusual Scaling Of Magnetotransport
S. Nair, M. Nicklas, F. Steglich, J.L. Sarrao, J.D. Thompson, A.J. Schofield and S. Wirth
Phys. Rev. B 79, 094501 (2009).
74. Divergence Of The Magnetic Grüneisen Ratio At The Field-Induced Quantum Critical Point In $YbRh_2Si_2$
Y. Tokiwa, T. Radu, C. Geibel, F. Steglich and P. Gegenwart
Phys. Rev. Lett. 102, 066401 (2009).
75. Kondo-Cluster-Glass State Near A Ferromagnetic Quantum Phase Transition
T. Westerkamp, M. Deppe, R. Küchler, M. Brando, C. Geibel, P. Gegenwart, A.P. Pikul and F. Steglich
Phys. Rev. Lett. 102, 206404 (2009).
76. Evidence For Re-Entrant Superconductivity In $EuFe_2As_2$ Under Pressure
C.F. Miclea, M. Nicklas, H.S. Jeevan, D. Kasinathan, Z. Hossain, H. Rosner, P. Gegenwart, C. Geibel and F. Steglich
Phys. Rev. B 79, 212509 (2009).
77. Electrical Resistivity Of $Yb(Rh_{1-x}Co_x)_2Si_2$ Single Crystals At Low Temperatures
S. Friedemann, N. Oeschler, C. Krellner, C. Geibel and F. Steglich
Physica B 404, 2890 (2009).

78. Analysis Of The Normal State Magnetotransport In $CeIrIn_5$
S. Nair, M. Nicklas, J.L. Sarrao, J.D. Thompson, F. Steglich and S. Wirth
J. Supercond. Nov. Magn. 22, 195 (2009).
79. Magnetotransport In The $CeIrIn_5$ System: The Influence Of Antiferromagnetic Fluctuations
S. Nair, M. Nicklas, J.L. Sarrao, J.D. Thompson, F. Steglich and S. Wirth
J. Supercond. Nov. Magn. 22, 201 (2009).
80. Phase Transition And Anomalous Low Temperature Ferromagnetic Phase In $Pr_{0.6}Sr_{0.4}MnO_3$ Single Crystals
S. Rößler, S. Harikrishnan, C.M. Naveen Kumar, H.L. Bhat, Suja Elizabeth, U.K. Rößler, F. Steglich and S. Wirth
J. Supercond. Nov. Magn. 22, 205 (2009).
81. Phase Transitions And Rare-Earth Magnetism In Hexagonal And Orthorhombic $DyMnO_3$ Single Crystals
S. Harikrishnan, S. Rößler, C.M. Naveen Kumar, H.L. Bhat, U.K. Rößler, S. Wirth, F. Steglich and S. Elizabeth
J. Phys.: Condens. Matter 21, 096002 (2009).
82. $FeSb_2$: Prototype Of Huge Electron-Diffusion Thermoelectricity
P. Sun, N. Oeschler, S. Johnsen, B.B. Iversen and F. Steglich
Phys. Rev. B 79, 153308 (2009).
83. Vortex Avalanches In The Noncentrosymmetric Superconductor Li_2Pt_3B
C.F. Miclea, A.C. Mota, M. Sigrist, F. Steglich, T.A. Sayles, B.J. Taylor, C. McElroy and M.B. Maple
Phys. Rev. B 80, 132502 (2009).
84. Thermoelectric Transport Properties Of Highly Oriented $FeSb_2$ Thin Films
Y. Sun, S. Johnsen, P. Eklund, M.B. Sillassen, J. Bøttiger, N. Oeschler, P. Sun, F. Steglich and B.B. Iversen
J. Appl. Phys. 106, 033710 (2009).
85. Thermopower Evidence For An Abrupt Fermi Surface Change At The Quantum Critical Point Of $YbRh_2Si_2$
S. Hartmann, N. Oeschler, C. Krellner, C. Geibel, S. Paschen and F. Steglich
Phys. Rev. Lett. 104, 096401 (2010).
86. Detaching The Antiferromagnetic Quantum Critical Point From The Fermi Surface Reconstruction In $YbRh_2Si_2$
S. Friedemann, T. Westerkamp, M. Brando, N. Oeschler, S. Wirth, P. Gegenwart, C. Krellner, C. Geibel and F. Steglich Nature Phys. 5, 465 (2009).
87. Pressure Induced Change Of The Pairing Symmetry In Superconducting $CeCu_2Si_2$
E. Lengyel, M. Nicklas, H.S. Jeevan, G. Sparn, C. Geibel, F. Steglich, Y. Yoshioka and K. Miyake
Phys. Rev. B 80, 140513(R) (2009).
88. Violation Of Critical Universality At The Antiferromagnetic Phase Transition Of $YbRh_2Si_2$

C. Krellner, S. Hartmann, A. Pikul, N. Oeschler, J. G. Donath, C. Geibel, F. Steglich and J. Wosnitza
Phys. Rev. Lett. 102, 196402 (2009).

89. Extreme Vortex Pinning In The Non-Centrosymmetric Superconductor $CePt_3Si$
C.F. Miclea, A.C. Mota, M. Nicklas, R. Cardoso, F. Steglich, M. Sigrist and E. Bauer
Phys. Rev. B 81, 014527 (2010).

90. Huge Thermoelectric Power Factor: $FeSb_2$ VERSUS $FeAs_2$ AND $RuSb_2$
P. Sun, N. Oeschler, S. Johnsen, B.B. Iversen and F. Steglich
Appl. Phys. Express 2, 091102 (2009).

91. Magnetism And Superconductivity Driven By Identical 4f States In A Heavy-Fermion Metal
S. Nair, O. Stockert, U. Witte, M. Nicklas, R. Schedler, K. Kiefer, J.D. Thompson, A.D. Bianchi, Z. Fisk, S. Wirth and F. Steglich
Proc. Natl. Acad. Sci. USA 107, 9537 (2010).

92. Superconductivity Versus Quantum Criticality: What Can We Learn From Heavy Fermions? (invited paper)
F. Steglich, J. Arndt, S. Friedemann, C. Krellner, Y. Tokiwa, T. Westerkamp, M. Brando, P. Gegenwart, C. Geibel, S. Wirth and O. Stockert
J. Phys.: Condens. Matter 22, 164202 (2010).

93. Characteristics Of The Magnetic Order in $CeCu_2Si_2$ Revealed By Neutron Spin-Echo Measurements
J. Arndt, O. Stockert, E. Faulhaber, P. Fouquet, H.S. Jeevan, C. Geibel, M. Loewenhaupt and F. Steglich
J. Phys.: Conference Series 200, 012009 (2010).

94. Quantum Griffiths Phase In $CePd_{1-x}Rh_x$ With $x \approx 0.8$
M. Brando, T. Westerkamp, M. Deppe, P. Gegenwart, C. Geibel and F. Steglich
J. Phys.: Conference Series 200, 012016 (2010).

95. Electron Scattering Off Structural Two-Level Systems In $ZrAs_{1.595}Se_{0.393}$
T. Cichorek, L. Bochenek, R. Niewa, M. Schmidt, A. Schlechte, R. Kniep and F. Steglich
J. Phys.: Conference Series 200, 012021 (2010).

96. Electron Spin Resonance Of The Itinerant Magnets $ZrZn_2$ And $Nb_{1-y}Fe_{2+y}$: A Comparison
T. Förster, J. Sichelschmidt, D. Grüner, M. Brando, N. Kimura and F. Steglich
J. Phys.: Conference Series 200, 012035 (2010).

97. Quantum Criticality In $Yb(Rh_{0.97}Co_{0.03})_2Si_2$ Probed By Low-Temperature Resistivity
S. Friedemann, N. Oeschler, C. Krellner, C. Geibel and F. Steglich
J. Phys.: Conference Series 200, 012038 (2010).

98. Thermopower Of $Ce_{1-x}La_xTiGe$ With Highly Degenerate Orbital Ground State
N. Oeschler, M. Deppe, S. Hartmann, N. Caroca-Canales, C. Geibel and F. Steglich
J. Phys.: Conference Series 200, 012148 (2010).

99. H-T Phase Diagram Of $YbCo_2Si_2$ With H II [100]
L. Pedrero, M. Brando, C. Klingner, C. Krellner, C. Geibel and F. Steglich
J. Phys.: Conference Series 200, 012157 (2010).

100. Interacting Magnetic Sublattices And Ferrimagnetism In Sr-Doped $DyMnO_3$

S. Rößler, S. Harikrishnan, U.K. Rößler, S. Elizabeth, H.L. Bhat, F. Steglich and S. Wirth
J. Phys.: Conference Series 200, 012168 (2010).

101. Magnetization Study On The Field-Induced Quantum Critical Point In $YbRh_2Si_2$
Y. Tokiwa, C. Geibel, F. Steglich and P. Gegenwart
J. Phys.: Conference Series 200, 012205 (2010).

102. Non-Magnetic Kondo Effect In M-As-Se (M = Zr, Hf, Th) Phases
T. Cichorek, L. Bochenek, Z. Henkie, M. Schmidt, A. Czulucki, R. Kniep and F. Steglich
Phys. Status Solidi B 247, 586 (2010).

103. NQR Study Of Pressure Induced Superconductivity In $CeCu_2Si_2$
K. Fujiwara, Y. Okazaki, Y. Hata, K. Miyoshi, J. Takeuchi, T.C. Kobayashi, C. Geibel and F. Steglich
Phys. Status Solidi B 247, 608 (2010).

104. Angular Dependence Of The Upper Critical Field B_{c2} In A/S Type $CeCu_2Si_2$
H.A. Vieyra, N. Oeschler, H.S. Jeevan, C. Geibel and F. Steglich
Phys. Status Solidi B 247, 611 (2010).

105. Scanning Tunneling Microscopy Studies On $CeCoIn_5$ And $CeIrIn_5$
S. Ernst, S. Wirth, F. Steglich, Z. Fisk, J.L. Sarrao and J.D. Thompson
Phys. Status Solidi B 247, 624 (2010).

106. Lack Of Magnetic Ordering In $Ce_{1-x}La_xNi_2Ge_2$
A. Pikul, C. Geibel, N. Oeschler, M.E. Macovei, N. Caroca-Canales and F. Steglich
Phys. Status Solidi B 247, 691 (2010).

107. The Crossed-Field And Single-Field Hall Effect In $LuRh_2Si_2$
S. Friedemann, N. Oeschler, S. Wirth, F. Steglich, S. MaQuilon and Z. Fisk
Phys. Status Solidi B 247, 723 (2010).

108. Effect Of Chemical Substitution And Pressure On $YbRh_2Si_2$
M. Nicklas, M. E. Macovei, J. Ferstl, C. Krellner, C. Geibel and F. Steglich
Phys. Status Solidi B 247, 727 (2010).

109. Development Of The Critical Exponent At The Antiferromagnetic Phase Transition Of $YbRh_2Si_2$ Under Chemical Pressure
C. Krellner, C. Klingner, C. Geibel and F. Steglich
Phys. Status Solidi B 247, 734 (2010).

110. Nuclear Contribution To The Specific Heat Of $Yb(Rh_{0.93}Co_{0.07})_2Si_2$
A. Steppke, M. Brando, N. Oeschler, C. Krellner, C. Geibel and F. Steglich
Phys. Status Solidi B 247, 737 (2010).

111. Ge-Based Skutterudites MPt_4Ge_{12}: A Comparative ^{195}Pt NMR Study
M. Baenitz, R. Sarkar, R. Gumeniuk, A. Leithe-Jasper, W. Schnelle, H. Rosner, U. Burkhardt, M. Schmidt, U. Schwarz, D. Kaczorowski, Y. Grin and F. Steglich
Phys. Status Solidi B 247, 740 (2010).

112. Low-Temperature Properties Of The Electron Spin Resonance In $YbRh_2Si_2$
J. Sichelschmidt, T. Kambe, I. Fazlishanov, D. Zakharov, H.-A. Krug von Nidda, J. Wykhoff, A. Skvortsova, S. Belov, A. Kutuzov, B.I. Kochelaev, V. Pashchenko, M. Lang, C. Krellner, C. Geibel and F. Steglich

Phys. Status Solidi B 247, 747 (2010).
113. Thermoelectric Properties Of The Clathrate I Ba_8Ge_{433}
C. Candolfi, U. Aydemir, M. Baitinger, N. Oeschler, F. Steglich and Yu. Grin
J. Electron. Mat. 39, 2039 (2010).
114. Pressure-Temperature Phase Diagram Of The Ferromagnetic Kondo Lattice Compound CeRuPO
M.E. Macovei, M. Nicklas, C. Krellner, C. Geibel and F. Steglich
Physica B 404, 2934 (2009).
115. FERMI-Surface Collapse And Dynamical Scaling Near A Quantum Critical Point
S. Friedemann, N. Oeschler, S. Wirth, C. Krellner, C. Geibel, F. Steglich, S. Paschen, S. Kirchner and Q. Si
Proc. Natl. Acad. Sci. USA 107, 14547 (2010).
116. Why Could Electron Spin Resonance Be Observed In A Heavy Fermion Kondo Lattice?
B.I. Kochelaev, S.I. Belov, A.M. Skvortsova, A.S. Kutuzov, J. Sichelschmidt, J. Wykhoff, C. Geibel and F. Steglich
Eur. Phys. J. B 72, 485 (2009).
117. Separation Of Energy Scales In Undoped $YbRh_2Si_2$ Under Hydrostatic Pressure
Y. Tokiwa, P. Gegenwart, C. Geibel and F. Steglich
J. Phys. Soc. Jpn. (Letters) 78, 123708 (2009).
118. Narrow Band Gap And Enhanced Thermoelectricity In $FeSb_2$
P. Sun, N. Oeschler, S. Johnsen, B.B. Iversen and F. Steglich
Dalton Transactions 39, 1012 (2010).
119. Evidence For A Non-Fermi-Liquid Phase In GE-Substituted $YbRh_2Si_2$
J. Custers, P. Gegenwart, C. Geibel, F. Steglich, P. Coleman and S. Paschen
Phys. Rev. Lett. 104, 186402 (2010).
120. Crystal Structure And Transport Properties Of $BaGe_{433}$
U. Aydemir, C. Candolfi, H. Borrmann, M. Baitinger, A. Ormeci, W. Carrillo-Cabrera, C. Chubilleau, B. Lenoir, A. Dauscher, N. Oeschler, F. Steglich and Y. Grin
Dalton Transactions 39, 1078 (2010).
121. Probing Quantum Criticality And Its Relationship With Superconductivity In Heavy Fermions (invited paper)
P. Gegenwart and F. Steglich
Developments in Quantum Phase Transitions, L.D. Carr, ed. (Taylor and Francis, LLC, London and Tokyo, 2010) (in press).
122. Magnetically Driven Superconductivity In $CeCu_2Si_2$
O. Stockert, J. Arndt, E. Faulhaber, C. Geibel, H.S. Jeevan, S. Kirchner, M. Loewenhaupt, K. Schmalzl, W. Schmidt, Q. Si and F. Steglich
Nature Phys. (in press).
123. Universal Signatures Of The Metamagnetic Quantum Critical Endpoint: $CeRu_2Si_2$
F. Weickert, M. Brando, F. Steglich, P. Gegenwart and M. Garst
Phys. Rev. B 81, 134438 (2010).
124. Anisotropic Electron Spin Resonance Of YbIr2Si2
T. Gruner, J. Wykhoff, J. Sichelschmidt, C. Krellner, C. Geibel and F. Steglich
J. Phys.: Condens. Matter 22, 135602 (2010).

125. Thermoelectric Power Of Correlated Compounds
N. Oeschler, S. Hartmann, U. Köhler, M. Deppe, P. Sun and F. Steglich
Properties and Applications of Thermoelectric Materials, V. Zlatić and A.C. Hewson, eds. (Springer Science + Business Media B.V., 2009) p. 81.
126. High Resolution Magnetostriction Measurements In Pulsed Magnetic Fields Using Fibre Bragg Gratings
R. Daou, F. Weickert, M. Nicklas and F. Steglich
Rev. Sci. Instrum. 81, 033909 (2010).
127. Pair Breaking By Nonmagnetic Impurities In The Noncentrosymmetric Superconductor $CePt_3Si$
M. Nicklas, F. Steglich, J. Knolle, I. Eremin, R. Lackner and E. Bauer
Phys. Rev. B 81, 180511(R) (2010).
128. Crystal Chemistry And Physical Properties Of The Nonmagnetic Kondo-Compound $HfAs_{1.7}Se_{0.2}$
A. Czulucki, G. Auffermann, M. Bednarski, L. Bochenek, M. Böhme, T. Cichorek, R. Niewa, N.Oeschler, M. Schmidt, F. Steglich and R. Kniep
Chem. Phys. Chem. 11, 2639 (2010).
129. Crystal Structure Investigations Of $ZrAs_xSe_y$ ($x > y$, $x + y \leq 2$) By Single Crystal Neutron Diffraction At 300 K, 25 K And 2.3 K
R. Niewa, A. Czulucki, M. Schmidt, G. Auffermann, T. Cichorek, M. Meven, B. Pedersen, F. Steglich and R. Kniep
J. Solid State Chem. 183, 1309 (2010).
130. Orientation Control And Thermoelectric Properties Of $FeSb_2$ Films
Y. Sun, E. Zhang, S. Johnsen, M. Sillassen, P. Sun, F. Steglich, J. Bottiger and B.B. Iversen
J. Phys. D: Appl. Phys. 43, 205402 (2010).
131. Effect Of Pressure On The Electron Spin Resonance Of A Heavy Fermion Metal
J. Sichelschmidt, J. Wykhoff, T. Gruner, C. Krellner, C. Klingner, C. Geibel, F. Steglich, H.-A. Krug von Nidda, D.V. Zakharov, A. Loidl and I. Fazlizhanov
Phys. Rev. B 81, 205116 (2010).
132. Evolution Of Magnetism In $Yb(Rh_{1-x}Co_x)_2Si_2$ Single Crystals
C. Klingner, C. Krellner, M. Brando, D.V. Vyalikh, K. Kummer, S. Danzenbächer, S.L. Molodtsov, T. Kinoshita, T. Muro, C. Geibel and F. Steglich
Phys. Rev. B (submitted).
133. Magnetic And Electronic Quantum Criticality In $YbRh_2Si_2$ (invited paper)
S. Friedemann, T. Westerkamp, M. Brando, N. Oeschler, P. Gegenwart, C. Krellner, C. Geibel, S. MaQuilon, Z. Fisk, F. Steglich and S. Wirth
J. Low Temp. Phys. 103, 67 (2010).
134. Divergence Of The Grüneisen Parameter And Magnetocaloric Effect At Heavy-Fermion Quantum Critical Points (invited paper)
P. Gegenwart, Y. Tokiwa, J.G. Donath, R. Küchler, C. Bergmann, H.S. Jeevan, E.D. Bauer, J.L. Sarrao, C. Geibel and F. Steglich
J. Low Temp. Phys. 103, 117 (2010).

135. Search For Electronic Phase Separation At Quantum Phase Transitions C. Pfleiderer, P. Böni, C. Franz, T. Keller, A. Neubauer, P.G. Niklowitz, P. Schmakat, M. Schulz, Y.-K. Huang, J.A. Mydosh, M. Vojta, W. Duncan, F.M. Grosche, M. Brando, M. Deppe, C. Geibel, F. Steglich, A. Krimmel and A. Loidl J. Low Temp. Phys. 103, 167 (2010).

136. Atomically Resolved Stm On Perovskite Manganite Single Crystals
S. Rössler, B. Padmanabhan, S. Elizabeth, H.L. Bhat, F. Steglich and S. Wirth
Appl. Phys. Lett. 96, 202512 (2010).

137. Hall Effect Measurements And Electronic Structure Calculations On $YbRh_2Si_2$ And ITS Reference Compounds $LuRh_2Si_2$ AND $YbIr_2Si_2$
S. Friedemann, S. Wirth, N. Oeschler, C. Krellner, C. Geibel, F. Steglich, S. MaQuilon, Z. Fisk, S. Paschen and G. Zwicknagl
Phys. Rev. B 82, 035103 (2010).

138. Heavy Fermions And Quantum Phase Transitions (invited paper)
Q. Si and F. Steglich
Science 319, 1161 (2010)

139. Low-Temperature Properties Of $Ceru_4sn_6$ From Nmr And Specific Heat Measurements: Heavy Fermions Emerging From A Kondo-Insulating State
E. M. Brüning, M. Brando, M. Baenitz, A. Bentien, A. M. Strydom, R. E. Walstedt and F. Steglich
Phys. Rev. B 82, 125115 (2010).

140. $BaGe_5$: A New Type Of Intermetallic Clathrate
U. Aydemir, L. Akselrud, W. Carrillo-Cabrera, C. Candolfi, N. Oeschler, M. Baitinger, F. Steglich and Y. Grin
J. Am. Chem. Soc. 132, 10984 (2010).

141. Heavy-Fermion Superconductivity (invited paper) F. Steglich In: Centennial Book on Superconductivity, P. Kes and H. Rogalla (eds.) (in press).

OTHER CONTRIBUTIONS OF INTEREST TO SUPERCONDUCTIVITY RESEARCH: PROJECTS, GRANTS ETC.

Member, Committee of Experts on Superconductivity, German Federal Ministry of Research and Technology, 1987-90

In: Superconductivity and Condensed Matter … ISBN 978-1-61209-785-5
Editor: Collin J. Eckles

Chapter 18

BIOGRAPHICAL SKETCH – Superconductivity Research

NAME:	TITLE:	DATE OF BIRTH:
SEUNG JOO LEE	PROFESSOR	09 MAY 1948

EDUCATION

Institution and Location	Degree	Year Conferred	Scientific Field
Korea Military Academy,	B.S.	1966-1970	Chemistry
Seoul	B.S.	1972-1974	Physics
Seoul National Univ., Seoul	M.S.	1975-1977	Physics
Seoul National Univ., Seoul	PhD	1977-1981	Physics
State Univ. of New York,			
Buffalo NY	Post Doc	1985-1986	Electrical Eng
Princeton University,			
Princeton NJ			

CONTACT POINTS

Quantum-functional Semiconductor Research center, Dongguk University, Seoul 100-715 Korea, Tel:82-2-2260-3953, Fax: 82-2-2260-3945, e-mail: leesj@dongguk.edu

RESEARCH AND PROFESSIONAL EXPERIENCE

Dr. Seung Joo Lee currently serves as Research Professor in the Quantum-functional Semiconductor Research Center (QSRC) of Dongguk University in Seoul, Korea (since Jan 2001). He has been a member of the American Physical Society and a Fellow of the Korean Physical Society since 1982. Dr. Lee's background is especially interesting in that he served on the faculty of the Korea Military Academy for nearly 25 years (March 1974-Dec. 2000) and retired as Dean of the Academy's Academic Board with the military rank of Brigadier General. He has made special contributions to increasing the knowledge of physics in the military community of Korea and worked to enhance the relationship between the military and civilian communities in the scientific arena. This significant service is in addition to his outstanding accomplishments as a scientist of international renown.

Dr. Lee has distinguished himself as a world-recognized expert in the field of condensed matter theory. His areas of accomplishments include semiconductor quantum well lasers, quantum dots, ferroelectric/ferromagnetic properties of films, low-dimensional electron gases, and semiconductor hetero-structure systems in the presence of a high magnetic field. He has also contributed to the development of the generalized random walk theory and its successful application to determining the carrier capture probability in a quantum well system. His work has led to over 110 research articles in leading journals (Physical Review Letters, Applied Physics Letters, Physical Review B, Physics Reports, and so on). He has also authored two books – one on general college physics and the other on quantum dots and quantum wells.

Dr. Lee provided outstanding service to the physics community during 2000-01 in his role as Vice President of the Korean Physical Society, where he also has had a four-year appointment (2008-12) to the International Affairs Committee. He has served as Chief Editor of the *Journal of the Korean Physical Society* and *The New Physics* and is working as a group member of the First Asia-Europe Physics Summit entitled "Physics towards Science Innovations" under the aegis of the European Physical Soceity and Association of the Asia-Pacific Physical Societies. His devotion to the international physics community has been shown through the Commission on Semiconductors of the International Union of Pure and Applied Physics where he has been serving two terms (2006-2008, and 2009-2011) and has brought the International Conference on Physics of Semiconductors-30 to Seoul Korea this year and has been serving as a program co-chair.

His contributions to physics and society have been recognized by a number of prestigious awards, including the Post Doctoral Fellowship of the Korea Science and Engineering Foundation (1985), a special Citation (1994), Order of National Service Merit (1997), and Order of National Security Service Merit (2000) from the President of the Republic of Korea, and the Medal for Distinguished Service (1995) from the Korean Military Academy.

Dr. Lee has expanded his horizons by serving in various positions overseas. These include visiting professorships/appointments at Princeton University, State University of New York at Buffalo, Max Planck Institute in Stuttgart, Wuerzburg University in Germany, and International Center for Theoretical Physics in Trieste, Italy. .

Professional Appointments

That is described above.

Honors relevant to research

Described as above.

PUBLICATIONS DURING LAST THREE YEARS

- "Spin-polarized current in a magnetic barrier structure under microwave radiation," K. C. Seo, G. Ihm, and S. J. Lee, " Physica Status Solidi (c), Volume4, Issue2, Date:February 2007, Pages:319-321.
- "Spin dependent Fano effect in an Aharonov-Bohm interferometer with a magnetic quantum dot", K. C. Seo, G. Ihm, and S. J. Lee, Physica Status Solidi (c), Volume4, Issue2, Date:February 2007, Pages:322-324.
- "Fano resonance in an Aharonov-Bohm interferometer with a quantum dot," K. C. Seo, G. Ihm, and S. J. Lee, Journal of Kor. Phy. Soc. (JKPS), 50 1647-1650 (2007).
- Effect of Thermal Annealing on the Magnetic and the Optical Properties of (Ga1-xMnx)N Thin Films Grown on GaN Buffer Layers, H. C. Jeon, Y. Shon, H. S. Kim, M. K. Li, S. J. Lee, T. W. Kang, J. H. Jung, T. W. Kim, H. S. Kim, M. D. Kim, K. L. Wang and J. J. Lee, JKPS, 2007 50:1921-1924.
- Structural, Optical, and Magnetic Characteristics of an InMnP:Zn Epilayer, Yoon Shon , H. C. Jeon , Sejoon Lee , Seung Joo Lee , T. W. Kang , Jin Soak Kim , Eun Kyu Kim , Chong S. Yoon and Jeoung Ju Lee, JKPS, 2007 50:814-818.
- Self-consistent solutions of the Curie temperature in ferromagnetic/ferroelectric hybrid double quantum wells, S. H. Park, S. W. Ryu, J. J. Kim, W. P. Hong, and H. M. Kim, N. Kim, J. W. Kim, T. W. Kang, S. J. Lee*, and S. N. Yi, Solid State Communication, *Volume 143, Issues 8-9, August 2007, Pages 395-398* (2007)
- Shifting intrinsic bistability states in a ferroelectric resonant tunneling diode M. K. Li, N. M. Kim, S. J. Lee, J. W. Kim, and T. W. Kang , Phys. Rev. B 75, 212106 (2007)
- Ferroelectric and magnetic properties of CdMnS films prepared by coevaporation D. H. Kim, D. J. Lee, N. M. Kim, S. J. Lee, T. W. Kang, Y. D. Woo, and D. J. Fu, J. Appl. Phys. 101, 094111 (2007)
- Polarization switching effects in ferromagnetic/ferroelectric hybrid double quantum wells, Nammee Kim, Seoung-Hwan Park, Jin Woo Kim, Tae Won Kang, Seung Joo Lee , and Gukhyung Ihm, AIP Conf. Proc. 893, 607 (2007)
- Optoelectronic Applications of ZnO/ZnMgO Quantum well Lasers in the Blue and the UV Spectral Regions, Seoung-Hwan Park , Doyeol Arm , Tae Won Kang , and Seung Joo Lee, AIP Conf. Proc. 893, 261 (2007)
- Spin dependent current in a modified Aharonov–Bohm interferometer, K.C. Seo, G. Ihm,_, S.J. Lee, Physica E: Low-dimensional Systems and Nanostructures, 40 (2008) 2185–2187
- Transitions of density of states and heat capacity in spherical quantum structure formed by a thin barrier, Seung Joo Lee, Hee Chang Jeon, Tae Won Kang, Satofumi Souma, Physica E: Low-dimensional Systems and Nanostructures, 40 (2008) 2198–2200
- Dimensional Transitions of Density of States and Heat Capacity in Spherical Quantum Dots, S. J. Lee*, H C Jeon, T W Kang, S. Souma, Int. J.of Modern Physics B, 22, (15) 2373 - 2382(2008)
- Electronic structure and magnetism in transition metals-doped 8-hydroxy-quinoline aluminum, J. M. Baik, Y. Shon, S. J. Lee, Y. H. Jeong, T. W.Kang J. L. Lee, J. American Chem Soc vol. 130 (41), pp 13522-13523 (2008)
- The Mass Effects on Two-dimensional Relativistic Fermions, M. K. Li, S. J. Lee*, T.W. Kang, Current Applied Physics 9 (2009) 769–772

- Theoretical Study on the Photocurrent of a ZnO-based Quantum Dot Photodetector, M. K. Li, S. J. Lee* and T. W. Kang, Journal of the Korean Physical Society, Vol. 55, No. 1, (July 2009), pp. 120-123
- Properties of laser gain of Li:CdZnO/ZnMgO quantum well, H.C. Jeon, S.J.Lee*, S.H.Park T.W.Kang, Physica B: Physics of Condensed Matter 404 (2009), pp. 4341-4343
- Optical Properties of Wurtzite ZnO-based Quantum Well Structures with Piezoelectric and Spontaneous Polarizations, Seoung-Hwan Park, Doyeol Ahn, Sam Nyung Yi, Tae Won Kang, and Seung Joo Lee*, Book Chapter of World Scientific, "Computational Studies of New Materials"(2009)
- Electronic and Optical Properties of CdZnO Quantum Well Structures with Electric Field and Polarization Effects, H.Jeon, Seoung-Hwan Park, Seung Joo Lee*, T. W. Kang, and Thomas F. George, . Applied Physics Letters 96, 101113 (2010 March)
- Transport in armchair graphene nanoribbons modulated by magnetic barriers, Nojoon Myounga, G. Ihm*, S. J. Lee, Physica E (2010) doi:10.1016/j.physe.2009.12.020
- nhancement of optical gain in CdZnO/ZnMgO quantum well lasers, Hee Change Jeon, Seung Joo Lee*, Seoung-Hwan Park, Tae Won Kang, Doyeol Ahn, and G. Ihm, Physica E (2010)doi:10.1016/j.physe.2009.12.052
- On the Specific Heat of Diluted Magnetic Semiconductor GaMnAs, M. K. Li, S. J. Lee et al Submitted (2010)

Other contributions of interest to research: Projects, Grants etc.

Described as above.

At Iguasu

In: Superconductivity and Condensed Matter ... ISBN 978-1-61209-785-5
Editor: Collin J. Eckles

Chapter 19

BIOGRAPHICAL SKETCH – Superconductivity Research

NAME:	TITLE:	DATE OF BIRTH:
SHI-DONG LIANG	PROFESSOR	NOVEMBER 20, 1958

EDUCATION

Institution and Location	Degree	Year Conferred	Scientific Field
1.Foshan University	1. Diploma	July 1981	Physics
2.Shanghai Institute of Nuclear Research, Academia Sinica	2. M. Phil.	July 1990	Physics
3. The University of Hong Kong	3. Ph. D.	Dec. 1999	Physics
4. Institute of Physics, Academia, Sinica, Taiwan	4.Postdoctoral research fellow	1999-2001	Physics

RESEARCH AND PROFESSIONAL EXPERIENCE

PROFESSIONAL APPOINTMENTS

Professor 2005—, Sun Yat-Sen University, Guangzhou, PR China;
Associate Professor 2002-2005, Sun Yat-Sen University, Guangzhou, PR China;
Postdoctoral research fellow, 1999-2001, Institute of Physics, Academia, Sinica, Taiwan
Lecturer 1994-1996, Guangzhou University, Guangzhou, PR China;
Lecturer 1992-1993, Foshan University, Foshan, P.R. China

HONORS RELEVANT TO RESEARCH

1997 The Award for Progress on Science and Technology of Guangdong Provice (Member)
Influence of phonon and external field on electron and exciton in quantum well
2007 The Award for Progress on Science and Technology of Guangdong Provice (Member)
Nano cold cathode and its device

PUBLICATIONS DURING LAST THREE YEARS

1. Shi-Dong Liang, Lu Chen, Theories of field and thermionic emission from carbon nanotubes, J. of Vac. Sci. and Tech. B28 (2), C2A50 (2010).
2. Guangyao Huang and Shi-Dong Liang, Orbital magnetic phase and pure spin persistent current in spin-orbit coupling mesoscopic rings, Europhysics letter, 86, (2009) 67009.
3. Shi-Dong Liang and Lu Chen, Generalized Fowler-Nordheim Theory of Field Emission of Carbon Nanotubes, Phys. Rev. Lett.101,027602(2008)
4. Shi-Dong Liang, Bardeen–Cooper–Schrieffer formalism of superconductivity in carbon nanotubes, Physica B403, 2288 (2008)
5. Shi-Dong Liang, Electronic structure and quantum phase transition in multi-wall carbon nanotubes, Int. J. Mod. Phys. B 21, 4377(2007)
6. Shi-Dong Liang, N.Y. Huang, S. Z. Deng and N.S. Xu, Quantum effect in the field emission of carbon nanotubes, J. of Vac. Sci. and Tech. B24, (2006).
7. Shi-Dong Liang, N.Y. Huang, S. Z. Deng and N.S. Xu, Intrinsic energy spectrum of single-wall carbon nanotubes in field-emission, Phys. Rev. B73.245301 (2006).
8. Shi-Dong Liang, Y.H. Bai, and B. Peng, Peierls instability and persistent current in mesoscopic conducting polymer rings, Phys. Rev. B74, 113304 (2006)
9. Shi-Dong Liang, S. Z. Deng and N.S. Xu, Seeking for an optimal performance of multi-wall carbon nanotubes in field-emission, Phys. Rev. B74, 155413 (2006).
10. H. Tang, Shi-Dong Liang, S. Z. Deng and N.S. Xu, Comparison of field and thermonic emission of single-wall carbon nanotubes, J. of Phys. D39, 5280(2006).

OTHER CONTRIBUTIONS OF INTEREST TO RESEARCH: PROJECTS, GRANTS ETC.

RESEARCH PROJECTS

1. 2004-2005, Guangdong Science and Technology Fund
   Physical mechanism of field emission from carbon nanotubes,
2. 2007-2010 National Science and Technology Fund
   Physical mechanism of spin polarized field emission based on nanomaterials

In: Superconductivity and Condensed Matter … ISBN 978-1-61209-785-5
Editor: Collin J. Eckles

Chapter 20

BIOGRAPHICAL SKETCH – Superconductivity Research

NAME:	TITLE:	DATE OF BIRTH:
CHIA-JYI LIU	PROFESSOR	SEP. 26, 1962

EDUCATION

Institution and Location	Degree	Year Conferred	Scientific Field
National Taiwan University, Taiwan	B. Sc.	1984	Organic Chemistry
The Jhons Hopkins University, U. S. A.	Ph. D.	1991	Solid State Physics and Chemistry
The Johns Hopkins University, U. S. A.	Postdoctor	1991 (6 months)	Solid State Physics and Chemistry
Southern Illinois University at Carbonale	Postdoctor	1991-1992	Electrodeposition of Superconductive Films

CONTACT POINTS

National Changhua Univeristy of Education, Department of phsyics, 1 jin de raod, changhua, 500 Taiwan, liucj@cc.ncue.edu.tw

RESEARCH AND PROFESSIONAL EXPERIENCE

PROFESSIONAL APPOINTMENTS:

- 1999-present Professor, National Changhua University of Education, Taiwan
- 2007 Visiting Professor, Argonne National Laboratory, U. S. A. (4 months)

- 2000 Visiting Professor, Industrial Research Limited, New Zealand (3 months)
- 1996-1999 Associate Professor, National Changhua University of Education, Taiwan
- 1996-1996 Research Scientist, Industrial Research Limited, New Zealand (3 months)
- 1995-1996 Research Fellow, Physics Department, Victoria University of Wellington, New Zealand
- 1992-1994 Visiting Scientist, Superconducting Research Laboratory, International Superconductivity Technology Center, Japan

HONORS RELEVANT TO SUPERCONDUCTIVITY RESEARCH

- NCUE Alumni Association Creative Invention Award, Taiwan (2009)
- 45th National Science Council Overseas Short-Term Research Fellow (2007)
- NCUE Alumni Association Academic Award, Taiwan (2007)
- NCUE Alumni Association Academic Award, Taiwan (2005)
- NCUE Alumni Association Academic Award, Taiwan (2000)
- 38th National Science Council Overseas Short-Term Research Fellow (2000)
- NCUE Alumni Association Academic Award, Taiwan (1999)
- Class A Research Award, National Science Council, Taiwan (1997).
- Technical Excellence Award, International Superconductivity Technology Center, Japan (1994).

PUBLICATIONS DURING LAST THREE YEARS

1. N. V. Nong, **Chia-Jyi Liu**, M. Ohtaki, 2010, "High-temperature thermoelectric properties of late rare earth-doped $Ca_3Co_4O_{9+\delta}$" , to appear in J. Alloy Compd.
2. **Chia-Jyi Liu**, Chia-Yuan Liao, Jiunn-Yuan Lin, and Jyh-Fu Lee, 2010, "Synthesis of mono-layer and bi-layer of cobalt oxyhydrates $(Na,K)_x(H_2O)_yCoO_{2-\square}$", Philosophical Magazine, 90, 1585-1597.
3. N. V. Nong, **Chia.-Jyi Liu**, M. Ohtaki, 2010, "Improvement on the high temperature thermoelectric performance of Ga-doped misfit-layered $Ca_3Co_{4-x}Ga_xO_{9+\delta}$ (x = 0, 0.05, 0.1, and 0.2)", 2010, J. Alloy and Compounds, 491, 53-56.
4. **Chia-Jyi Liu**, Shu-Yo Chen, Long-Jiann Shih, Hsueh-Jung Huang, 2010, "Fabrication of nanotubules of thermoelectric γ-$Na_{0.7}CoO_2$ using porous aluminum oxide membrane as supporting template", Mater. Chem. and Phys. 119, 424-427.
5. Yow-Jon Lin, Mu-Shan Wang, **Chia-Jyi Liu**, Hsueh-Jung Huang, 2010, "Defects, stress and abnormal shift of the (0 0 2) diffraction peak for Li-doped ZnO films", Appl. Surf. Sci., 256, 7623-7627.
6. Chia-Lung Tsai, Mu-Shan Wang, Ya-Hui Chen, Hsing-Cheng Chang, **Chia-Jyi Liu**, Ching-Ting Lee, Yu-Tai Shih, Hsueh-Jung Huang, and Yow-Jon Lin, 2010 " Effects of Li content on the structural, optical, and electrical properties of LiZnMgO films", J. Appl. Phys. 107, 113717-113720.

7. **Chia-Jyi Liu**, Gao-Jhih Liu, Chun-Wei Tsao, Yo-Jhih Huang, 2009, "Improvement of thermoelectric power factor of hydrothermally prepared $Bi_{0.5}Sb_{1.5}Te_3$ compared to its solvothermally prepared counterpart", J. Electronic Materials, 38, 1499.
8. J. L. Chen, Y. S. Liu, **Chia-Jyi Liu**, L.-C. Huang, C. L. Dong, S. S. Chen, C. L. Chang, 2009, "Effect of Mn doping on the physical properties of misfit-layered $Ca_3Co_4O_{9+\square}$", accepted for publication in J. Phys. D: Appl. Phys.
9. **Chia-Jyi Liu**, Pradipta K. Nayak, and Yong-Zhi Chen, 2009, "A simple approach of fabricating thermoelectric γ-Na_xCoO_2 and superconductive $Na_x(H_2O)_yCoO_{2-\square}$ films using the sol-gel spin coating method", accepted for publication in Thin Solid Films.
10. **Chia-Jyi Liu,** Pradipta K.Nayak,Zhi-Ru Lin ,Kai-Yi Jeng,2008, "The effects of sintering temperature on preparation, resistivity,and thermopower of c-axis oriented $Ca_3Co_{3.95}Fe_{0.05}O_{9+\square}$ films fabricated using sol-gel spin coating method", Thin Solid Films, 516, 8564.
11. Yow-Jon Lin,Ping-Hsun Wu,Chia-Lung Tsai, **Chia-Jyi Liu,** Ching-Ting Lee, Hsing-Cheng Chang,Zhi-Ru Linand Kai-Yi Jeng, 2008, "Mechanisms of enhancing band-edge liuminescence of $Zn_{1-X}Mg_XO$ prepared by the sol-gel method", J. Phys. D: Appl. Phys., 41, 125103.
12. Yow-Jon Lin, Wei-Chung Chen, Yi-Min Chin, and **Chia-Jyi Liu**, 2009, "Hysteresis mechanism in current–voltage characteristics of ZrO_x films prepared by the sol–gel method, J. Phys. D: Appl. Phys. 42, 045419.
13. J Y Lin, Y L Chen, Ben Hsu, C J Liu, Y L Tai, C A Hsieh and J F Lee, 2009, "Effects of water intercalation in $Na_xCoO_2{\cdot}yH_2O$", Journal of Physics: Conference Series, 150, 052138.
14. **Chia-Jyi Liu,** Pradipta K.Nayak,Zhi-Ru Lin ,Kai-Yi Jeng, 2008, "The effects of sintering temperature on preparation, resistivity,and thermopower of c-axis oriented $Ca_3Co_{3.95}Fe_{0.05}O_{9+\square}$ films fabricated using sol-gel spin coating method", Thin Solid Films, 516, 8564.
15. Yow-Jon Lin,Ping-Hsun Wu,Chia-Lung Tsai, **Chia-Jyi Liu,** Ching-Ting Lee, Hsing-Cheng Chang,Zhi-Ru Linand Kai-Yi Jeng, 2008, "Mechanisms of enhancing band-edge liuminescence of $Zn_{1-X}Mg_XO$ prepared by the sol-gel method", J.Phys. D: Appl. Phys., 41, 125103.
16. Yow-Jon Lin,Chia-Lung Tsai,Wei-Chung Chen, **Chia-Jyi Liu,** Lance Horng, Yu-Tai Shih, Zhi-Ru Lin, Jia-Feng Wang, 2008, "Mechanisms of enhancing magnetic properties of $Zn_{1-X}Co_XO$ films prepared by the sol-gel method", Journal of Crystal Growth, 310, 3763.
17. Yow-Jon Lin,Ping-Hsun Wu,Chia-Lung Tsai, **Chia-Jyi Liu,** Zhi-Ru Lin, Hsing-Cheng Chang,and Ching-ting Lee, 2008, "Effects of Mg incorporation on the optical properties of ZnO prepared by the sol-gel method", J. Appl. Phys., 103, 113709.
18. **Chia-Jyi Liu,** Chia- Yuan Liao ,Wen-Ching Hung , Jung-Sheng ang, Hwo-Shuenn Shue, 2008, "Water intercalation of sodium cobalt oxides using aqueous permanganate solution", Journal of Physics and Chemistry of Solids, 69, 1483.
19. Yow-Jon Lin , Wei-Chung Chen, Hsing-Cheng Chang, **Chia-Jyi Liu,** Zhi-Ru Lin, 2008, "Analysis of the band-edge luminescence degradation for ZnO films with Al doping prepared by the sol-gel method", Journal of Crystal Growth", 310, 4110.

20. **Chia-Jyi Liu**, Chia-Yuan Liao, Jung-Sheng Wang, Pradipta K. Nayak, Zhi-Ru Lin, Hwo-Shuenn Sheu, 2007, "Thermopower and Co K-edge studies of potassium sodium cobalt oxyhydrates $Na_{0.33}K_{0.02}(H_2O)_{1.33}CoO_{2-\square}$ and $Na_{0.07}K_{0.21}(H_2O)_{0.63}CoO_{2-\square}$", Appl. Phys. Lett., 91, 142509.
21. **Chia-Jyi Liu**, 2007, Comment on "Characterization of Sodium Cobalt Oxides Related to P3-Phase Superconductor", Chem. Mater., 19, 5200.
22. Y.-J. Chen, **C.-J. Liu**, J.-S. Wang, J.-Y. Lin, C. P. Sun, S. W. Huang, J. M. Lee, J. M. Chen, J. F. Lee, D. G. Liu, H. D. Yang, 2007, "Effect of Mn impurities on the superconductivity in $Na_xCoO_2 \cdot yH_2O$", Phys. Rev. B, 76, 092501.
23. **Chia-Jyi Liu**, Pradipta K. Nayak, G. V. M. Willaims, 2007, "Magnetothermopower and Magnetoresistivity of $RuSr_2Gd_{2-x}Ce_xCu_2O_{10+\square}$ (x=0.6, 1.0)", Appl. Phys. Lett., 91, 123110.
24. **Chia-Jyi Liu**, Tsung-Hsien Wu, Lin-Li Hsu, Jung-Sheng Wang, Shu-Yo Chen, 2007, "Synthesis of superconducting cobalt oxyhydrates using a novel method: Electrolyzed and oxidized water", Physica C, 460-462, 493.
25. **Chia-Jyi Liu**, Jeng-Lung Chen, Li-Chen Huang, Zhi-Ru Lin and Ching-Lin Chang, 2007, "X-ray absorption spectroscopy studies of Fe-doped misfit-layered $Ca_3Co_{4-x}Fe_xO_{9+\square}$ (x = 0, 0.05, 0.1, and 0.15)", J. Appl. Phys., 102, 014908.
26. Chia-Lung Tsai, Yow-Jon Lin, Ping-Hsun Wu, Shu-You Chen, Day-Shan Liu, Jia-Huang Hong, **Chia-Jyi Liu**, Yu-Tai Shih, Jie-Min Cheng, Hsing-Cheng Chang, 2007, "Induced increase in surface work function and surface energy of indium tin oxide-doped ZnO films by $(NH_4)_2S_x$ treatment", J. Appl. Phys., 101, 113713.
27. J.-Y. Lin, Y.-J. Chen, **C.-J. Liu**, J.-S. Wang, C.P. Sun, H. D. Yang, 2007, "Specific heat of Mn-doped $Na_xCoO_2 \cdot yH_2O$", Physica C, 460-462, 475.
28. J.-Y. Lin, Y.-J. Chen, **C.-J. Liu**, J.-S. Wang, C.P. Sun, H. D. Yang, 2007, "Induced magnetic moment by Mn doping in $Na_xCo_{1-z}Mn_zO_2 \cdot yH_2O$", J. Magn. Magn. Mater., 310, e312.

ISBN 978-1-61209-785-5

Chapter 21

BIOGRAPHICAL SKETCH – Superconductivity Research

NAME:	TITLE:	DATE OF BIRTH:
CHIH WEI LUO	ASSOCIATE PROFESSOR	1975/11/4

EDUCATION:

Institution and Location	Degree	Year Conferred	Scientific Field
National Chiao Tung University	B.S.	1998	Electrophysics
National Chiao Tung University	Ph.D.	2003	Electrophysics, Ultrafast dynamics phenomena
National Chiao Tung University	Postdoctoral Researcher	2004	Electrophysics, Ultrafast dynamics phenomena, Ultrashort pulse optics

RESEARCH AND PROFESSIONAL EXPERIENCE:

- January 2004 – June 2004
  Special Researcher, Industrial Technology Research Institute, Taiwan.
- September 2002 – June 2003
  Visiting research at Humboldt University and Max-Born Institute in Berlin by sandwich program of NSC in Taiwan.
- September 1999 – June 2000
  Teacher of Special Topics of Physics, National Experimental High School at the Science-Based Industrial Park.

PROFESSIONAL APPOINTMENTS

- April/2006 – present
  Assistant Professor, Department of Electrophysics, National Chiao Tung University Hsinchu, Taiwan

HONORS RELEVANT TO SUPERCONDUCTIVITY RESEARCH:

- 2009 **Youth Innovation Papers Award,** The 7th international conference on new theories, discoveries and applications of superconductors and related materials (New^3-SC 2009)

PUBLICATIONS DURING LAST THREE YEARS

[2010]

1. **C. W. Luo***, H. P. Lo, C. H. Su, I. S. Wu, Y.-J. Chen, K. H. Wu, J.-Y. Lin, T. M. Uen, J. Y. Juang, and T. Kobayashi,
   "*Doping dependence of the ultrafast electronic dynamics of* $Y_{1-x}Pr_xBa_2Cu_3O_7$ *thin-film superconductors from femtosecond optical spectroscopy*",
   Phys. Rev. **B *in Press*** (2010).
2. **C. W. Luo***, Y.-J. Chen, T. C. Huang, K. H. Wu, J.-Y. Lin, J. Y. Juang, and T. M. Uen,
   "*Electron-boson coupling constant of oxygendeficient* $Y_{0.7}Ca_{0.3}Ba_2Cu_3O_{7-\delta}$ *measured by ultrafast spectroscopy*",
   J. Supercond. Nov. Magn. ***in Press*** (2010).
3. **C. W. Luo***
   "*Ultrafast quasiparticle dynamics in heavy Ca-doped YBCO thin films*",
   Physica C, ***in Press*** (2010).
4. **C. W. Luo***,
   "*Observation of vortex-like excitations in Pr-doped YBCO thin films from femtosecond spectroscopy*",
   Physica C, ***in Press*** (2010).
5. **C. W. Luo***, T. C. Huang, M. T. Chiang, K. H. Wu, J. Y. Juang, J.-Y. Lin, and T. M. Uen,
   "*Anisotropic ultrafast dynamics of quasiparticles on* CuO_2 *planes of* $Y_{0.7}Ca_{0.3}Ba_2Cu_3O_{7-\delta}$",
   J. Supercond. Nov. Magn. **23,** 781 (2010).
6. K. H. Wu*, I. C. Gou, **C. W. Luo**, T. M. Uen, J.-Y. Lin, J.Y. Juang, C. K. Chen, J. M. Lee and J. M. Chen,
   "*Anisotropic electronic structure in single crystalline orthorhombic* $TbMnO_3$ *Thin Films*",
   Thin Solid Films 5**18,** 2275 (2010).

[2009]

7. T. H. Lin*, C. C. Hsieh, **C. W. Luo**, J.-Y. Lin, C. P. Sun, H. D. Yang, C.-H. Hsu, Y. H. Chu, K. H. Wu, T. M. Uen, and J. Y. Juang*,
   "*Magnetism-induced ferroelectric polarization in the c-axis-oriented orthorhombic HoMnO3 thin films*",

J. Appl. Phys. **106,** 103923 (2009).
8. J W Chou, K C Lin, Y T Tang, F K Hsueh, Yao-Jen Lee, **C. W. Luo**, Y N Chen, C T Yuan, Hsun-Chuan Shih, W C Fan, M C Lin, W C Chou, and Der-San Chuu*, "*Observation of the Localized Surface Plasmons in Spatially Controlled Array Structures*", Nanotechnology **20**, 415201 (2009).
9. H. C. Shih*, T. H. Lin, **C. W. Luo***, J.-Y. Lin, T. M. Uen, J. Y. Juang, K. H. Wu, J. M. Lee, J. M. Chen, and T. Kobayashi, "*Magnetization dynamics and Mn^{3+} d-d excitation in hexagonal $HoMnO_3$ revealed by wavelength-tunable time-resolved femtosecond spectroscopy*", Phys. Rev. **B 80**, 024427 (2009)
✦It has been selected for the August 2009 issue of Virtual Journal of Ultrafast Science
10. J. W. Chou, Kao-Chin Lin, Yao-Jen Lee, Chi-Tsu Yuan, Fu-Kuo Hsueh, Hsun-Chuan Shih, Wen-Chung Fan, **Chih-Wei Luo**, Ming-Chieh Lin, Wu-Ching Chou, and Der-San Chuu*, "*Observation of the Localized Surface Plasmons in Spatially Controlled Array Structures*", Nanotechnology **20**, 305202 (2009).
11. **C. W. Luo***, Y. T. Wang, F. W. Chen, H. C. Shih, and T. Kobayashi, "*Eliminate coherence spike in reflection-type pump-probe measurements*", Opt. Express **17**, 11321 (2009).
✦It has been selected for the August 2009 issue of Virtual Journal of Ultrafast Science
12. K. H. Wu*, T. Y. Hsu, H. C. Shih, Y. J. Chen, **C. W. Luo**, T. M. Uen, J.-Y. Lin, J.Y. Juang, and T. Kobayashi, "*Ultrafast Optical Probes of Polaron Dynamics in $La_{0.7}Ca_{0.3}MnO_3$ Thin Films*", J. Appl. Phys. **105,** 043901 (2009).
13. T. H. Lin*, H. C. Shih, C. C. Hsieh, **C. W. Luo**, J.-Y. Lin, J. L. Her, H. D. Yang, K. H. Wu, T. M. Uen, and J. Y. Juang, "*Strain-induced effects on magnetic ordering and magneto-capacitance in orthorhombic HoMnO3 thin films*", J Phys-Condens Mat **21**, 026013 (2009)
14. Y. C. Tsai*, W. J. Chang, C. C. Hsieh, **C. W. Luo**, K. H. Wu, T. M. Uen, J. Y. Juang, and J.-Y. Lin, "*Phase separation and persistent magnetic memory effect in $La_{0.625}Ca_{0.375}MnO_3$ and $La_{0.375}Pr_{0.25}Ca_{0.375}MnO_3$ films*", J. Appl. Phys. **105**, 013705 (2009).

[2008]
15. **C. W. Luo***, C. C. Lee, C. H. Li, H. C. Shih, Y.-J. Chen, C. C. Hsieh, C. H. Su, W. Y. Tzeng, K. H.Wu, J. Y. Juang, T. M. Uen, S. P. Chen, J.-Y. Lin, and T. Kobayashi, "*Ordered YBCO sub-micron array structures induced by pulsed femtosecond laser irradiation*", Opt. Express **16**, 20610 (2008).
16. C. S. Yang*, K. F. Chien, J. Y. Lai, **C. W. Luo**, W. C. Chou, Y. T. Shih, J. S. Wang, and S. R. Jian, "*Carrier Dynamics in Self-Assembled CdTe Stranski-Krastanow Quantum Dots Grown on ZnSe by Molecular Beam Epitaxy*", J. Korean Phys. Soc. **53**, 2905 (2008)
17. C. C. Hsieh*, T. H. Lin, H. C. Shih, C.-H. Hsu, **C. W. Luo**, J.-Y. Lin, K. H. Wu, T. M. Uen, and J. Y. Juang,

"*Magnetic ordering anisotropy in epitaxial orthorhombic multiferroic* $YMnO_3$ *films*" J. Appl. Phys. **104**, 103912 (2008).

18. S.-A. Ku, **C. W. Luo**, H.-L. Lio, K.-H. Wu, J.-Y. Juang, A. I. Potekaev, O. P. Tolbanov*, S. Yu. Sarkisov, Yu. M. Andreev, and G. V. Lanskii, "*Optical properties of nonlinear solid solution* $GaSe_{1-x}S_x$ $(0<x\leq 0.4)$ *crystals*", Russ. Phys. J. **51**, 1083 (2008).
19. Bo-An Tsai, Yao-Jen Lee*, Hsin-Yi Peng, Pei-Jer Tzenf, **Chih-Wei Luo**, and Kuei-Shu Chang-Liao, "*Using Spike-Anneal to Reduce Interfacial Layer Thickness and Leakage Current in Metal-Oxide-Semiconductor Device with TaN/Atomic Layer Deposition-Grown HfAlO/Chemical Oxide/Si Structure*", Jpn. J. Appl. Phys. **47**, 2438 (2008).
20. T. H. Lin*, C. C. Hsieh, H. C. Shih, **C. W. Luo**, T. M. Uen, K. H. Wu, J. Y. Juang*, J.-Y. Lin, C.-H. Hsu, and S. J. Liu, "*Anomalous magnetic ordering in b-axis-oriented orthorhombic* $HoMnO_3$ *thin films*", Appl. Phys. Lett. **92**, 132503 (2008).
21. **C. W. Luo***, Y. Q. Yang, I. T. Mak, Y. H. Chang, K. H. Wu, and T. Kobayashi, "*A widely tunable dual-wavelength CW Ti:sapphire laser with collinear output*" Opt. Express **16**, 3305 (2008).

In: Superconductivity and Condensed Matter … ISBN 978-1-61209-785-5
Editor: Collin J. Eckles

Chapter 22

BIOGRAPHICAL SKETCH – Superconductivity Research

NAME:	TITLE:	DATE OF BIRTH:
RAGAB M. MAHANI	DR.	24/12/1966

EDUCATION:

Institution and Location	Degree	Year Conferred	Scientific Field
Helwan university – Egypt	B.Sc Degree	1991	Experimental Physics
Helwan university – Egypt	M.Sc	1999	Experimental Physics
Rostock University - Germany	PhD	2005	Experimental Physics

RESEARCH AND PROFESSIONAL EXPERIENCE

Feb. 2006: Permanent position as Researcher, National Research Center, Cairo, Egypt.
May 2005: PhD Degree of Physics, Rostock University, Germany.
1995: - M.Sc of Physics, Faculty of Science, Helwan University, Cairo, Egypt.
Permanent position as Researcher Assistant, National Research Center, Microwave Physics and Dielectric Dep., Cairo, Egypt.
1993: Permanent position as Assistant Researcher, National Research Center, Microwave Physics and Dielectric Dep., Cairo, Egypt.
1992 – 1993: Pre-master study, Faculty of Science, Helwan University, Cairo, Egypt.
1991: B.Sc Degree of physics (Very Good 81 %.), Faculty of Science, Helwan University, Cairo, Egypt.

1995 – till now: Member of Egyptian Society of Polymer Science and Technology, National Research Center, Cairo, Egypt.

Professional Appointments

I am utilizing the dielectric spectroscopy phenomenon to study the dynamic molecular behavior for the following:

- Polymers
- Conventional and advanced ceramics
- Breast Cancerous tissues
- Water Hyacinth Plant

Good experience for the following techniques

- Broadband Dielectric Spectrometer BDS4000 (Alpha high resolution dielectric analyzer). Also, Hewlett - Packard impedance Analyzer HP4284 A. Both techniques were used for measuring dielectric properties of polymeric materials.
- Differential Scanning Calorimetric; PerkinElmer DSC6, to study thermal properties of different of polymeric materials.
- Small Angle X-ray Scattering (SAXS): at the European Synchrotron Radiation Facility (ESRF) in Grenoble, France, to study the morphology during crystallization of polymers.

Honors relevant to scientific research

June 2010: Prize in Physics (Dr. Shahata Farag's Prize), National Research Center, Cairo Egypt.

Publications During Last Three Years

1- The Open Spectroscopy Journal, The Open Spectroscopy Journal, 2010, 4, 32-40.
Effect of Physical and Chemical Treatments on the Electrical and Structural Properties of Water Hyacinth
2- Industerial Ceramics Vol. 30, 1/2010
Physico-electrical properties of ceramic bodies based on nepheline tailing and clay
3- Journal Of alloys and Compounds – 2010 – In press
Structure and dielectric behavior of nano-structure ferroelectric Bax Sr1-x TiO3 prepared by sol - gel method.
4- Carbohydrate Polymers 80 (2010) 367–373

Study mechanical, swelling and dielectric properties of prehydrolysed banana fiber – Waste polyurethane foam composites

5- Microelectronic Engineering 87 (2010) 2218–2224
Structural, electrical and magnetic characterizations of Ni/Cu/p-Si Schottky diodes prepared by liquid phase epitaxy

6- Physica B 404 (2009) 2274–2279
Structure and dielectric studies of nano-composite Fe_2O_3: $BaTiO_3$ preparedby sol–gel – method

7- American Ceramic Society Bulletin (2008), Vol. 87, No. 4
The addition of B_2O_3 or LiF to electroceramic bodies with increased quartz esulted in porcelains that can be used as insulating materials in the frequency range 1 kHz to 1 MHz

8- Egypt, J. of Appl. Sci., 22(4A) 2007
Dielectric Properties of Ferroelectric Nano-composite Ba_x Sr_{1-x} Tio_3 prepared by sol-Gel method.

PROJECTS

- Swedish Project Management in Multicultural Groups

Title: Synthesis of physical properties of nano - structure ferroelectric $BaSrTiO_3$

GRANTS

June 2009: Post Doc. In France for two weeks during the period of time between 15/06/2009 – 30/06/2009

In: Superconductivity and Condensed Matter …
Editor: Collin J. Eckles
ISBN 978-1-61209-785-5

Chapter 23

BIOGRAPHICAL SKETCH – Superconductivity Research

NAME:	TITLE:	DATE OF BIRTH:
HIROYUKI OYANAGI	PH. D	21/10/1951

EDUCATION

Institution and Location	Degree	Year Conferred	Scientific Field
University of Tokyo	Ph. D.	1980	x-ray spectroscopy
University of Tokyo	M.S.	1977	Material synthesis
University of Tokyo	B.S.	1975	Semiconductor

RESEARCH AND PROFESSIONAL EXPERIENCE

PROFESSIONAL APPOINTMENTS

April, 2008-Present — Visiting Professor, Science and Technology University of China

April, 2007 – April, 2008 — Professor, Ritsumeikan University

April, 2000 - Present — Principal Research Scientist, Photonics Research Institute, AIST

April, 2000 - April 2003 — Professor, Kyoto University

April, 1997 - March 2000	Associate Director, Physical Science Division, Electrotechnical Laboratory
April, 1992 - March 1997	Chief of Exotic Matter Physics Section,
April, 1987 - March 1992	Senior Researcher, Electrotecnical Laboratory
April, 1980 - March 1987	Researcher, Electrotecnical Laboratory

OCCUPATIONAL EXPERIENCE

Research subjects distribute over a wide range of materials. The goal of research efforts has been "local structure and function" or efforts to understand properties and functions based on short range order. Experience of synchrotron radiation instrumentation includes, stations using bending magnet radiation (BL10B, 11C, 4C/PF), multipole wiggler (BL13/PF) and undulator (BL10XU/Spring-8). Throughput the carrier, efforts have been focused on fluorescence detected x-ray absorption spectroscopy, enabling dilute and thin specimens to be studied in-situ. The initial research period focused on doped polymer metals, SiGe superlattices and III-V alloys and superlattices. A series of multiple solid state detector were developed including the most rcent pixel array detctor (PAD). Along with the R&D of x-ray detectors, efforts were made in the design and construction of the high power beamline using a multipole wiggler . Using the high flux beam and UHV apparatus with a MBE chamber, in-situ XAS experiments were achieved. After establishing the tuning undulator at Spring-8, the research was focused on photo-induced phenomena and presently laser-induced photostructural changes are the reseach topics. Since the beginning of high temperature superconductivity, local structure studies using XAS were also actively taken up in collaboration with University of Rome. Recent collaboration with Science and Technical University is also actively performed.

HONORS RELEVANT TO SUPERCONDUCTIVITY RESEARCH

May 1992	Director's Award of Science and Technology Agency

PUBLICATIONS DURING LAST THREE YEARS

2010

(1) C.J. Zhang, H. Oyanagi, Z.H. Sun, Y. Kamihara and H. Hosono, Electronic and lattice structures in $SmFeAsO_{1-x}F_x$ probed by x-ray absorption spectroscopy, PRB**81**, 094516 (2010).
(2) Z.H. Sun, H. Oyanagi, H. Nakamura, Y. Jiang, L. Zhang, M. Uehara, K. Yamashita, A. Fukano and H. Maeda, Ligand Effects of Amine on the Initial Nucleation and Growth Processes of CdSe Nanocrystals, J. Phys. Chem.C**114**, 10126 (2010).

(3) H. Oyangi, C.J. Zhang, Lattice Anomalies and HTSC in Pnictides and Cuprates Studied by XAS; Ploaron Resonance as a Common Clue, Journal of Superconductivity and Novel Magnetism (in press).
(4) H. Oyanagi, Z.H. Sun, Y. Jiang, M. Uehara, H. Nakamura, K. Yamashita, L. Zhang, C. Lee, A. Fukano and H. Maeda, In-situ XAFS spectrometer using a microfluidic cell: application to initial growth of CdSe nanocrystals, J. of Syn. Rad. (in press).

2009

(5) Zhihu Sun, Hiroyuki Oyanagi, Noriyuki Uchida, Takehide Miyazaki and Toshihiko Kanayama, Structure determination of W-capsulated Si cage clusters by x-ray absorption fine structure spectra, J. Phys. D: Appl. Phys. **42**, 015412 (2009).
(6) H. Oyangi, C.Zhang, A. Tsukada and M. Naito, Lattice Instability in High-Temperature Superconducting Cuprates: Polarons Probed by EXAFS, J. Supercond. Nov. Magn. **22**, 165, (2009).
(7) C.J. Zhang, and H. Oyanagi, Local lattice instability and superconductivity in $La_{1.85}Sr_{0.15}Cu_{1-x}M_xO_4$ (M=Mn, Ni, and Co), PRB **79**, 064521 (2009).
(8) Jian Ye, Yong Jiang, Qinghua Liu, Tao Yao, Zhiyun Pan, H. Oyanagi, Zhihu Sun, Wensheng Yan and Shiqiang Wei, Cosputtered Mn-doped si thin films studied by x-ray spectroscopy, J. of Appl. Phys. **106**, 103517 (2009).
(9) M. Uehara, Z.H. Sun, H. Oyanagi, K. Yamashita, A. Fukano, H. Nakamura and H. Maeda, In situ extended x-ray absorption fine structure study of initial processes in CdSe nanocrystals formation using a microreactor, Appl. Phys. Lett. **94**, 063104 (2009).
(10) Fengchun Hu, Zhiyun Pan, Tao Yao, Hiroyuki Oyanagi, Yong Jiang, Wensheng Yan, Bo He, Jian Ye and shiqiang Wei, Solubility of co-coped Cu ions in $Zn_{0.95}Co_{0.05}O$ diluted magnetic semiconductors, J. Phys. D: Appl. Phys, **42**, 125408,(2009).
(11) Z.H. Sun, H. Oyanagi, M. Uehara, H. Nakamura, K. Yamashita, a. Fukano and H. Maeda, Study on Initial Kinetics of CdSe Nanocrystas by a Combination of in Situ X-ray Absorption Fine Structure and Microfluidic Reactor, J. Phys. Chem. C **113**, 18608 (2009).
(12) H. Oyanagi, C.J. Zhang, Z.H. Sun, Y. Kamihara and H. Hosono, Low-temperature lattice anomaly in $LaFeAsO_{0.93}F_{0.07}$ probed by x-ray absorption spectroscopy: Evidence for a strong electron-phonon interaction, J. of Physics: Conference Series 190, 012094 (2009).
(13) M. Filippi, B. Kundys, S. Agrestini, W. Prellier, H. Oyanagi and N.L. Saini, Charge order, dielectric response, and local structure of $La_{5/3}Sr_{1/3}NiO_4$ system, J. of Appl. Phys. **106**, 104116 (2009).

2008

(14) C.J. Zhang, H. Oyanagi, Z.H. Sun, Y. Kamihara and H. Hosono, Low-temperature lattice structure anomaly in the $LaFeAsO_{0.93}F_{0.07}$ superconductor by x-ray absorption spectroscopy: Evidence for a strong electron-phonon interaction, PRB **78**, 214513(2008)
(15) H.M. Tsai, S.C. Ray, C.W. Pao, J.W. Chiou, C.L. Huang, C.H. Du, W.F. Pong, M.H. Tsai A. Fukano and H.Oyanagi, Enhancement of Si-O hybridization in low-temperature grown ultraviolet photo-oxidized SiO_2 film observed by x-ray absorption and photoemission spectroscopy, J. of App. Phys. **103**, 013704 (2008).
(16) H. Oyangi, C.Zhang, A. Tsukada and M. Naito, Lattice instability in high temperature superconducting cuprates probed by x-ray absorption spectroscopy, Journal of Physics: Conference Series **108**, 012038, (2008).

(17) C.J. Zhang and H. Oyanagi, The synthesis condition and its influence on Tc in Mn doped $La_{1.85}Sr_{0.15}CuO_4$, Physica C **468**, 1155 (2008)
(18) Z. Sun, W. Yan, G. Zhang, H. Oyanagi, Z. Wu, Q. Liu, W. Wu, T. Shi, Z. Pan, P. Xu, S. Wei, Evidence of substitutional Co ion clusters in $Zn_{1-x}Co_xO$ dilute magnetic semiconductors, Phys. Rev. B 77, 245208 (2008).
(19) C.J. Zhang, H. Oyanagi and C.H. Lee, Single crystal growth and characterization of $La_{2-x}Sr_xCuO_4$ with Mn doping, Physica C **468**, 898 (2008).
(20) H. Oyanagi, A. Tsukada, and M. Naito, Local Structure Study of T' Cuprate Superconductors, Journal of Physics and Chemistry of Solids 69, 2307 (2008).
(21) C.J. Zhang and H. Oyanagi, Effects of local lattice disorder on the superconductivity studied by XAFS, Journal of Physics: Conference Series 108, 012048 (2008).

2007

(22) C.J. Zhang, H. Oyanagi, B.H. Kim, Y.W. Park and Y.H. Zhang, Local lattice structure in Mn-doped $La_{2-x}Sr_xCuO_4$ studied by Cu and Mn *K*-edge XAFS, Phys. Rev.B **75**, 174504 (2007).
(23) H. Oyanagi, A. Tsukada, M. Naito, and N. L. Saini, Local structure of superconducting $(La,Sr)_2CuO_4$ under strain: Microscopic mechanism of strain-induced Tc variation, Phys. Rev. B **75**, 024511 (2007).

OTHER CONTRIBUTIONS OF INTEREST TO SUPERCONDUCTIVITY RESEARCH: PROJECTS, GRANTS ETC.

1. Chairman, International Symposium on Physics in Local Lattice Distortions (LLD2K)
July 23-26, 2000
Proceedings: Physics in Local Lattice Distortions: Fundamentals and Novel Concepts, AIP Conference Proceedings 554
American Institute of Physics
ISBN 1-56396-984-X
2. Chaiman, International Workshop on Electron States and Lattice Effects in Cuprate High Temperature Superconductors-Revealing True Story
Oct. 27-28, 2005
3. Chaiman, International Workshop on Electron States and Lattice Effects in Cuprate High Temperature Superconductors-Revealing True Story
Oct. 27-28, 2005

In: Superconductivity and Condensed Matter … ISBN 978-1-61209-785-5
Editor: Collin J. Eckles

Chapter 24

BIOGRAPHICAL SKETCH – Superconductivity Research

NAME:	TITLE:	DATE OF BIRTH:
ALEXANDER PANICH	DR.	JULY 5, 1949

EDUCATION

Institution and Location	Degree	Year Conferred	Scientific Field
Donetsk State University, Donetsk, USSR.	M. Sc.	1971	Physics
Institute of Physics, Siberian Branch of the Academy of Sciences of the USSR, Krasnoyarsk, USSR	Ph.D.	1981	Physics

CONTACT POINTS

Department of Physics, Ben-Gurion University of the Negev, P.O.Box 653, Be'er-Sheva 84105, Israel, e-mail: pan@bgu.ac.il

RESEARCH AND PROFESSIONAL EXPERIENCE

EMPLOYMENT HISTORY

1993-present Senior Research Scientist, KAMEA Fellow, Magnetic Resonance

	Laboratory, Department of Physics, Ben-Gurion University of the Negev, Be'er-Sheva, Israel
1997 (May-July)	Visiting Scientist, DAAD Fellow, Institute for Experimental Physics, Free University, Berlin, Germany.
1992-1993	Researcher, Infodisk Ltd., Advanced Technologies Center, Temed Industrial Park, Israel
1988- 1992	Senior Research Scientist, Department of Materials for Microelectronics, Institute of Inorganic Chemistry, Siberian Branch of the Academy of Sciences of the USSR, Novosibirsk, USSR.
1974-1988	Research Scientist, Department of Solid State Physics, Institute of InorganicChemistry, Siberian Branch of the Academy of Sciences of the USSR, Novosibirsk, USSR.
1971-1974	Engineer, Designer's Office, Institute of Inorganic Chemistry, Siberian Branch of the Academy of Sciences of the USSR, Novosibirsk, USSR.

EXPERIENCE

Nuclear Magnetic Resonance (NMR) and Nuclear Quadrupole Resonance (NQR) study of

(1) Nanomaterials (Nanodiamonds, Carbon onions, Carbon and Inorganic Fullerenes and Nanotubes, etc.)
(2) Low-dimensional solids,
(3) Incommensurate states,
(4) Graphite Intercalation Compounds,
(5) Inclusion Compounds,
(6) Phase transitions

PROFESSIONAL APPOINTMENTS

1999 – 2006 - Expert-Evaluator of INTAS proposals.

EDITORIAL BOARD: The Open Condensed Matter Physics Journal

Reviewer of scientific journals:

Physical Review B
Physical Review Letters
Journal of Physics: Condensed Matter
Journal of Physics D: Applied Physics
New Journal of Physics
Journal of Applied Physics

Journal of Physical Chemistry
Solid State Communications
The European Physical Journal B
Chemistry of Materials
Carbon
Journal of Low Temperature Physics
The Open Condensed Matter Physics Journal
Journal of Magnetic Resonance
Applied Magnetic Resonance
Nanotechnology
Semiconductor Science and Technology
Solid State Electronics
Physica Status Solidi
Materials Research Bulletin
Journal of Materials Science & Technology
Journal of Solid State Chemistry
Philosophical Magazine
Journal of Mathematical Chemisrty
Energy and Fuel
Journal of Fluorine Chemistry

HONORS RELEVANT TO RESEARCH

1966 - Gold Medal in electronics for schoolchildren, Exhibition of National Achievements of the USSR, Moscow.

1975, 1979 - Awards of the Young Scientists Conferences (Novosibirsk, Russia)

1981 - Academician Nikolaev's Scientific Conference Award (Novosibirsk, Russia)

1998 (15th Edition), 2001 (18th Edition), 2008 (25th Edition) - Who's Who in the World.

2006-2007 (9th Edition), 2008-2009 (10th Edition) - Who's Who in Science and Engineering.

PUBLICATIONS DURING LAST THREE YEARS

(a) Monographs

A. M. Panich, R. M. Sardarly, "Physical Properties of the Low-Dimensional A^3B^6 and $A^3B^3C_2^6$ Compounds". *Nova Science Publishers, Inc.,* New York, USA, 2010, 287 p.

(b) Articles in peer-reviewed scientific journals (2007-2010)

A. M. Panich, N. A. Sergeev, and I. Shlimak. Long-lived spin echoes in a magnetically dilute system: An NMR study of Ge single crystals. *Physical Review B* **76**, 155201/1-155201/8 (2007).

A.I. Shames, **A. M. Panich**, S.Porro, M.Rovere, S.Musso, A. Tagliaferro, M.V. Baidakova, V.Yu. Osipov, A.Ya. Vul', T. Enoki, M. Takahashi, E. Osawa, O.A.Williams, P.Bruno, D.M.Gruen.

Defects localization and nature in bulk and thin film utrananocrystalline diamond. *Diamond and Related Materials* **16**(10), 1806-1812 (2007).

A. M. Panich. Solid State Nuclear Magnetic Resonance Studies of NanoCarbons. (Review article). *Diamond and Related Materials* **16**(12), 2044-2049 (2007).

A. I. Shames, **A. M. Panich**, E. Mogilko, J. Grinblat, E.W. Prilutskiy, and E.A. Katz. Magnetic Resonance Study of Fullerene-Like Glassy Carbon. *Diamond and Related Materials* **16**(12), 2039-2043 (2007).

G. Cunningham, **A. M. Panich**, A. I. Shames, I. Petrov, O. Shenderova. Ozone-modified Detonation Nanodiamonds. *Diamond and Related Materials* **17** (4-5), 650-654 (2008).

T. Nakajima, T. Achiha, Y. Ohzawa, **A.M. Panich**, A.I. Shames. Electrochemical behavior of surface-fluorinated graphite in propylene carbonate-containing solvent. *Journal of Physics and Chemistry of Solids* **69**, 1292–1295 (2008).

A. M. Panich. Electronic properties and phase transitions in low-dimensional semiconductors (Topical Review). *J. Phys.: Condens. Matter* 20(29), 293202/1-42 (2008).

A. M. Panich, N. A. Sergeev, I. Shlimak, V. I. Ozhogin, and A.V. Tikhomirov. NMR study of the isotopically engineered Ge single crystals. *Hyperfine Interactions* **180**(1-3), 1-5 (2008).

A. M. Panich, S. Kashida, Single-Crystal NMR for the Layered Semiconductor $TlGaSe_2$ *J. Phys.: Condens. Matter* **20**(39) 395211/1-8 (2008).

A. I. Shames, E. A. Katz, **A. M. Panich**, D. Mogilyansky, E. Mogilko, J. Grinblat, V. P. Belousov, I. M. Belousova, and A. N. Ponomarev, Structural and Magnetic Resonance Study of Astralen Nanoparticles. *Diamonds and Related Materials* **18**, 505-510 (2009).

A. M. Panich, A. I. Shames, R. Rosentsveig, R. Tenne, Magnetic resonance study of MoS_2 fullerene-like nanoparticles. *J. Phys. Condensed Matter* **21** (39), 395301/1-6 (2009).

A. M. Panich, A.I. Shames, O. Medvedev, V.Yu. Osipov, A. E. Alexenskii, A. Ya.Vul', Magnetic Resonance Study of Detonation Nanodiamonds with Surface Chemically Modified by Transition Metal Ions. *Applied Magnetic Resonance* 36, 317-329 (2009).

Alexander M. Panich, Hans-Martin Vieth, Alexander I. Shames, Natalya Froumin, Eiji Ōsawa, and Akifumi Yao, Structure and bonding in fluorinated nanodiamond. J. Phys. Chem. C 114, 774-782 (2010).

A. I. Shames, **A. M. Panich**, V. Yu. Osipov, A. E. Aleksenskiy, A. Ya. Vul', T. Enoki, and K. Takai, Structure and Magnetic Properties of Detonation Nanodiamond Chemically Modified by Copper. *J. Applied Physics* **107**, 014318/1-10 (2010).

A. M. Panich, N. A. Sergeev, Nuclear spin-lattice relaxation in carbon nanostructures. *Physica B* **405**(8), 2034-2038 (2010).

A. I. Shames, I. Felner, V. Yu. Osipov, E. A. Katz, E. Mogilko, J. Grinblat, **A. M. Panich**, V. P. Belousov, I. M. Belousova, and A. N. Ponomarev. Closed π–electron Network in

Large Polyhedral Multi-shell Carbon Nanoparticles (*accepted by Nanoscience and Nanothechnology Letters*).

OTHER CONTRIBUTIONS OF INTEREST TO RESEARCH: PROJECTS, GRANTS ETC.

RESEARCH GRANTS:

Grants of Ministry of Science of Israel: 1994-1996, 1998-2000, 2009-2010.
Grants of the Israel Science Foundation: 1996-1999,
Seed Grants of Ben-Gurion University: 1998-1999, 1999-2000, 2001-2002, 2002-2003.

In: Superconductivity and Condensed Matter …
Editor: Collin J. Eckles
ISBN 978-1-61209-785-5

Chapter 25

BIOGRAPHICAL SKETCH – Superconductivity Research

NAME:	TITLE:	DATE OF BIRTH:
KONSTANTIN POPOV	DR.	NOVEMBER 2, 1949

EDUCATION

Institution and Location	Degree	Year Conferred	Scientific Field
Sankt-Peterburg State University	Diploma in Physics	1974	Physics. Thesis: Investigation of the afterglow of Ar plasma.
Post graduate correspondence course in Sankt-Peterburg State University		1977-1981	Spectroscopy
Sankt-Peterburg State University	PhD	1983	Physics. Thesis: Investigation of the broadening of the sodium and rubidium resonant lines by interaction with atoms and molecules
Sankt-Peterburg State University, Faculty of qualification level raising.	Certificate	1985	Computer Sciences
Syktyvcar State University, Faculty of pedagogical qualification level raising.	Diploma	1984 - 1985	Pedagogical sciences.

Institution and Location	Degree	Year Conferred	Scientific Field
Special course at Moscow State University	Certificate	1990	Physics. Raman scattering.
Special course at St.Peterburg State University	Certificate	1995	Physics. Photon correlation spectroskopy
Special course at St.Peterburg Nucliar physics Institute RAS	Certificate	2005	Computational physics

CONTACT POINTS

Laboratory of Theoretical and Computational Physics.
Department of Mathematics.
Komi Sciences Center, Ural Branch of Russian Academy of Science
24, Kommunisticheskaya str.
Syktyvkar, Komi, Russia, 167982,
Ph: +7-(8212)246242;
Fax: +7-(8212)215740
Email: kpopov@dm.komisc.ru

RESEARCH AND PROFESSIONAL EXPERIENCE

PROFESSIONAL APPOINTMENTS

Komi Sciences Center, Ural Branch, Russian Academy of Science, Department of Mathematics, 2002-till now.

LEADING RESEARCH ASSOCIATE

RESEARCH INTERESTS:

Computer modeling of quantum liquids. Investigation of strongly correlated Fermi systems. Development of the theory of fermion condensation. Calculation of the low temperature properties of such materials as metals with heavy fermions, superconductors, 2D liquid 3He.

Syktyvkar Institute of Chemistry, Komi Sciences Center, Ural Branch, Russian Academy of Science, Analytical laboratory, 1998-2002.

SENIOR RESEARCH ASSOCIATE

RESEARCH INTERESTS:

Application of Density Functional Theory (DFT) and Hartree-Fock (HF) Theory to calculations of atoms, molecules and small particles.

Syktyvkar State University, Syktyvkar, Russia, Department of Physics, Chair of Solid State Physics, 1974-2004.

ASSOCIATE PROFESSOR

PEDAGOGICAL DUTIES

Lectures, practical and laboratory lessons for optics, atomic physics and theoretical mechanics, advanced courses in Models of Amorphous Solids, lectures and practical lessons on Java programming.

HEAD OF THE EDUCATIONAL LABORATORY OF ATOMIC PHYSICS

RESEARCH INTERESTS

Structure and properties of gel (sol-gel transition, waves in gel media, nanocrystalline and fractal structure of gel, modeling of gel); metal-organic crystal growth in gel (nucleation, crystal growth, and crystal properties); Application of precision mercury measurements in geology (search of oil and gold deposits) and environmental control.

Heart brain structure: application of percolation theory to modeling of Brain structure, investigation of Electro encephalographic (EEG) patterns by means of non-line dynamics.

Valdosta State University, Valdosta, Ga, USA, Department of Physics, Astronomy and Geology, Fall quarter, 1997

VISITING PROFESSOR. Lectures in Physics Sciences

HONORS RELEVANT TO RESEARCH

Academic status: Docent, 1992

Society membership:

Physical Society of Russia (1991)
International Union of Crystallographers (1998)
Chemical Society of Russia (1998)
Ceramic Society of Russia (1998)
Mathematical Society of Cankt-Peterburg(2002)

PUBLICATIONS DURING LAST THREE YEARS

1. Shaginyan V. R., Popov K. G.V., Asymmetric tunneling, Andreev reflection and dynamic conductance spectra in strongly correlated metals. // Physics Letters A, 2007, v.361, p.406-412.
2. Shaginyan V. R., Popov K. G., Stephanovich V. A., E.V. Kirichenko E. V.V. Asymmetrical tunneling in heavy fermion metals as a possible probe for their non-Fermi liquid peculiarities. // Journal of Alloys and Compounds, 2007, v.442, p.29-33.
3. Shaginyan V.R., Popov K.G., Artamonov S.A.V. Universal Behavior of $CePd_{\{1-x\}}Rh_x$ Ferromagnet at the Quantum Critical Point. // JETP Letters, 2007, v.85, №8, p.398-403.
4. V.R. Shaginyan, M.Ya. Amusia, K.G. Popov, Universal Behavior of Strongly Correlated Fermi Systems, Usp. Fiz. Nauk 177 (2007) 586-618, [Phys. Usp. 50 (2007) 563-595.].
5. Shaginyan V. R., Popov K. G., Stephanovich V. A. Universal low-temperature behavior of $CePd_{\{1-x\}}Rh_x$ ferromagnet. // Europhys. Lett. 2007, v.79, p.47001-47006.

6. Shaginyan V. R., Popov K. G., Stephanovich V. A.\/., Universal Behavior of Two-Dimensional ^{3}He at Low Temperatures. // Phys.Rev.Lett.2008, v.100, p.096406-096409.
7. V.R. Shaginyan, P.G. Popov, S.A. Artamonov, General Propertes of a Magnetic-Field-Induced Landau Fermi Liquid in a High-Temperature Superconductors and Heavy fermion Metals, JETP Lett. 82 (2008) 215-219.
8. Shaginyan V.R., Amusia M. Ya., Popov K.G., Artamonov S.A.\/., Energy scales and the non-Fermi liquid behavior in $YbRh_2Si_2$. // JETP Lett., 2009, v.90, p.47-54.
9. Popov K.G., Shaginyan V.R., Fermi condansate phase transitiov cardinally change the Fermi liquid. //PersT, http://issp.ras.ru/Control/Inform/perst/ \\ 9_07/index.htm, 16, 7, 2009 г.
10. Shaginyan V.R., Popov K.G., General properties of $CePd_{1-x}Rh_x$ at quantum critical point. // Physica B: Physics of Condensed Matter, 2009, v.404, p.3179-3182.
11. Shaginyan V. R., Amusia M. Ya., Popov K. G., Stephanovich V. A., Quantum critical point in high-temperature superconductors. // Physics Letters A, 2009, v.373, p.686-692.
12. Shaginyan V. R., Amusia M. Ya., Msezane A. Z., Popov K.G., Stephanovich V. A.\/., Energy scales and magnetoresistance at a quantum critical point. // Physics Letters A, 373, v.2009, p.986-991.
13. Shaginyan V. R., Amusia M. Ya., Popov K.G.\/., Strongly correlated Fermi-systems: Non-Fermi liquid behavior, quasiparticle effective mass and their interplay. // Physics Letters A, 2009, v.373, p.2281-2286.
14. Shaginyan V. R., Msezane A. Z., Popov K.G., Stephanovich V. A.\/., Strongly correlated Fermi-systems: Magnetic-field-induced reentrance of Fermi-liquid behavior and spin-lattice relaxation rates in $YbCu_{5-x}Au_x$. // Physics Letters A, 2009, v.373, p.3783-3786.
15. Shaginyan V. R., Amusia M. Ya., Popov K.G.\/., Behavior of the antiferromagnetic phase transition near the fermion condensation quantum phase transition in $YbRh_2Si_2$. // Physics Letters A, 2010, v.374, v.659-665.
16. Shaginyan V. R., Popov K.G.\/., Fermion condensation: a strange idea successfully explaining behavior of numerous objects in Nature. // Ukrainian Journal of Physics, 2010, v.55, p.65-79.
17. Shaginyan V. R., Amusia M. Ya., Msezane A. Z., Popov K.G., Scaling behavior of heavy-fermion metals. // Physics Reports, 2010, v.492, p.31-109.

Other contributions of interest to research: Projects, Grants etc.

Grants

1. Grants of Ural Division of Academy of Science, Project: Creation of supercomputer laboratory of Mathematics Department of Komi Science Center. (2004-2010)
2. Research Grants of Russian Scientific Foundation. Universal properteis of strongly correlated fermi systems. (2009-1011).High resolution x-ray diffraction from composite, nanoporous and deformed structures. (2009-2010).
3. Research Grant of Komi Government. Syktyvkar State University. Project: Using of mercury measurement in geology (search of oil deposits) of Komi. (1996-1998).
4. Research Grants of Ministry of Higher Education of Russia. Syktyvkar State University. Project: Investigation of the crystal growth processes in gel. (1990-1995),

(1996-2000).

5. Research Grant of Ministry of Higher Education of Russia. Syktyvkar State University. Project: The structure and property of gel. (1994-1998).
6. Research Grant of The State Scientific Program of Russia: "Fundamental Problems of the Modern Physical Optics and It's Applications", Ministry of Higher Education of Russia. Syktyvkar State University. Project: Investigation of the non-line effects in spectra of media containing Small particles. 1993-1996,
7. Research Grant of Syktyvkar State University Project: Monitoring of mercury content in atmosphere air in Syktyvkar city. (1997-1998).

SCIENTIFIC EXPEDITIONS

1. 1983-1984. Search of diamond deposits on the Near-Secular-Polar Ural.
2. 1985-1986. Search and prospecting of deposits of porcelain stones of the Near-Secular-Polar Ural
3. 1988. Investigation of the Radio-Waves propagation. North Timan mountains. Russia.
4. 1996. Search and prospecting of the gas-oil deposits by means of using Mercury measurements at Ukhta region of Komi republic, Russia.
5. 1996. Search and prospecting of gold deposits by means of using mercury measurements at Secular-Polar Ural region of Komi republic, Russia.
6. 1997. Environmental investigations of mercury in the region of Kirkenes, Norway.
7. 1998. Monitoring of mercury content in the atmosphere air of Syktyvkar City.

In: Superconductivity and Condensed Matter …
Editor: Collin J. Eckles
ISBN 978-1-61209-785-5

Chapter 26

BIOGRAPHICAL SKETCH – Superconductivity Research

NAME:	DATE OF BIRTH:
JORGE QUINTANILLA	9 MAY 1974

EDUCATION

Institution and Location	Degree	Year Conferred	Scientific Field
University of Salamanca (Spain)	Licenciado	1997	Physics
University of Bristol (U.K.)	PhD	2002	Physics

RESEARCH AND PROFESSIONAL EXPERIENCE

PROFESSIONAL APPOINTMENTS

2002 — FAPESP Postdoc
University of Sao Paulo (USP) at Sao Carlos Brazil

2003 - 2004 — Leverhulme Research Fellow
University of Birmigham U.K.

2005 – 2009 — ATLAS Research Fellow in Condensed Matter Theory
ISIS Facility, STFC Rutherford Appleton Lab. U.K.

2010- — Lecturer and SEPNET Fellow
University of Kent at Canterbury U.K.
and ISIS Facility, STFC Rutherford Appleton Lab.

Honors relevant to research

1997	Awarded an Individual (Cat. 20) Marie Curie Fellowship
2005	Elected Fellow of St. Catherine's College
2009	Honorary Research Associate, University College London

Publications during last three years

Journal papers: refereed

[P13] S. T. Carr, J. Quintanilla and J. J. Betouras.
" Lifshitz transitions and crystallization of fully polarized dipolar fermions in an anisotropic two-dimensional lattice."
Phys. Rev. B **82**, 045110 (2010).
http://dx.doi.org/10.1103/PhysRevB.82.045110

[P12] O. Ciftja and J. Quintanilla.
"Effective Interaction Potentials in the Uppermost Landau Level."
Journal of Low Temperature Physics **159**, 189-192 (2010).
http://dx.doi.org/10.1007/s10909-009-0123-5

[P11] * J. Quintanilla, S. T. Carr and J. J. Betouras.
"Meta-nematic, smectic and crystalline phases of dipolar fermions in an optical lattice."
Physical Review A **79**, 031601**(R)** (2009).
http://dx.doi.org/10.1103/PhysRevA.79.031601

[P10] † A. D. Hillier, J. Quintanilla and R. Cywinski.
"Evidence for time-reversal symmetry breaking in the non-centrosymmetric superconductor $LaNiC_2$."
Physical Review Letters **102**, 117007 (2009).
http://dx.doi.org/10.1103/PhysRevLett.102.117007

[P9] * J. Quintanilla, K. Capelle and L. N. Oliveira.
"Density functional description of superconducting and magnetic proximity effects across a tunnelling barrier."
Physical Review B **78**, 205426 (2008). [Also in Virtual Journal of Applications of Superconductivity.]
http://dx.doi.org/10.1103/PhysRevB.78.205426

[P8] * J. Quintanilla, M. Haque and A. J. Schofield.
"Symmetry-breaking Fermi surface deformations from central interactions in two dimensions."
Physical Review B **78**, 035131 (2008).
http://dx.doi.org/10.1103/PhysRevB.78.035131

[P7] § V. L. Campo, K. Capelle, J. Quintanilla and C. Hooley.
"Quantitative determination of the Hubbard model phase diagram from optical lattice experiments by two-parameter scaling."
Physical Review Letters **99**, 240403 (2007).
http://dx.doi.org/10.1103/PhysRevLett.99.240403

CONFERENCE PAPERS: REFEREED

[C6] S. T. Carr, J. Quintanilla and J. J. Betouras.
"Deconfinement and quantum liquid crystalline states of dipolar fermions in optical lattices."
International Journal of Modern Physics B **23**, 4074-4086 (2009).
Invited paper, proceedings of CMT32.
http://dx.doi.org/10.1142/S0217979209063262

[C5] V. L. Campo, Jr., J. Quintanilla and C. Hooley.
"Possible critical behavior driven by the confining potential in optical lattices with ultra-cold fermions."
Physica B **404**, 3328-3331 (2009).
Contributed paper, proceedings of SCES'08.
http://dx.doi.org/10.1016/j.physb.2009.07.089

[C4] * J. Quintanilla, C. Hooley, B. J. Powell, A. J. Schofield and M. Haque.
"Pomeranchuk instability: symmetry breaking and experimental signatures."
Physica B **403**, 1279-1281 (2008).
Contributed paper, proceedings of SCES'07.
http://dx.doi.org/10.1016/j.physb.2007.10.126

OTHER PAPERS

Papers in the popular literature

[L1] * J. Quintanilla and C. Hooley.
"The strong-correlations puzzle."
Physics World **22**, 32-37 (June 2009).
http://physicsworldarchive.iop.org/index.cfm?action=summary&doc=22%2F06%2Fphwv22i06a38%40pwa-xml&qt=

Research Highlights

[H2] A. D. Hillier, J. Quintanilla, and R. Cywinski.
"Probing the symmetry of the non-centrosymmetric superconductor $LaNiC_2$."
ISIS 2009: The ISIS Facility Annual Report (Science and Technology Facilities Council, 2009).

OTHER CONTRIBUTIONS OF INTEREST TO RESEARCH: PROJECTS, GRANTS ETC.

Organiser, First SEPNet Summer Program in Condensed Matter Physics "Condensed Matter Physics in the City" (London, Egham and Harwell, July 2010).

Organiser, Strong Correlations Symposium, CMMP'09 (Warwick, December 2009).

2010 Principal investigator, £1450 University of Kent Strategic Research Development Grant "to examine the Bose-Hubbard model with quantum Monte Carlo to simulate two-parameter scaling."

2010 Principal investigator, 3-year SEPNet PhD studentship: "Harnessing novel excitations in strongly correlated matter: the road from magnetic monopoles to magnetronics."

In: Superconductivity and Condensed Matter …
Editor: Collin J. Eckles
ISBN 978-1-61209-785-5

Chapter 27

BIOGRAPHICAL SKETCH – Superconductivity Research

NAME:	TITLE:
DR. ASOK KUMAR RAY	PROFESSOR

EDUCATION

Institution and Location	Degree	Year Conferred	Scientific Field
Calcutta University, India	Bachelor of Science	1967	Physics
Calcutta University, India	Bachelor of Technology	1969	Radio-Physics & Electronics
Oklahoma State University, Stillwater, Oklahoma	Master of Science	1973	Physics
Texas Tech University, Lubbock, Texas	Master of Science	1975	Mathematics
Texas Tech University, Lubbock, Texas	Doctor of Philosophy	1977	Physics

CONTACT POINTS

Professor
Department of Physics
University of Texas at Arlington
Arlington, Texas 76019
Email: akr@uta.edu
Phone: 817-272-2503
817-272-2266

Research and Professional Experience

Academic Appointments

1971-1973	Graduate Teaching Assistant, Department of Physics, Oklahoma State University
1973-1974	Part-Time Instructor, Department of Mathematics, Texas Tech University
1974-1978	Instructor, Department of Physics, Texas Tech University
1977-1978	Research Assistant, Department of Mathematics, Texas Tech University
1978-1979	Visiting Assistant Professor, Department of Physics, Texas Tech University
1979-1981	Postdoctoral Research Associate, Quantum Theory Project, Departments of Physics and Chemistry, University of Florida
1981-1982	Assistant Professor, Department of Physics, Michigan Technological University
1979-1986	Staff Member, Institute for Numerical Transport Theory, Department of Mathematics, Texas Tech University
1982-1984	Visiting Assistant Professor, Department of Physics, University of Texas at Arlington
1984-1988	Assistant Professor, Department of Physics, University of Texas at Arlington
Summer, 1987	Visiting Research Fellow, University of East Anglia, Norwich, United Kingdom
1988 – 1992	Associate Professor, Department of Physics, University of Texas at Arlington
Fall, 1990	Guest Professor, Institute for Technical Electrochemistry, Technical University of Vienna, Austria
March, 1991	Visiting Research Fellow, Department of Theoretical Chemistry, Oxford University, United Kingdom
1992-cont.	Professor, Department of Physics, University of Texas at Arlington
August, 1998	Visiting Research Chemist, Department of Chemistry, Princeton University, Princeton, New Jersey
Oct.-Nov. 1998	Visiting Scientist, Los Alamos National Laboratory, Los Alamos, New Mexico
Dec. 1998-1999	Consultant, Los Alamos National Laboratory, Los Alamos, New Mexico
Dec. 1998-Jan. 1999	Visiting Research Chemist, Department of Chemistry, Princeton University, Princeton, New Jersey
July-Aug. 1999	Visiting Research Fellow, Department of Chemistry, Princeton University, Princeton, New Jersey.

Research Interests

Electronic Structures of the Actinides
Computational Electronic Structure Theory of Solids and Surfaces
Computational Studies of Nanostructures and Clusters
Computational Electron Transport Theory
Parallel Computing

Refereed Research Publications

1. *Invited Paper*: "Model-Independent Determination of Nuclear Radii," (with Y. N. Kim and S. Wald), Proceedings of the Second Nuclear Conference of the European Physical Society, Vol. II, 34-52, Cracow, Poland (1976).
2. "On the Effectiveness of the Inverse Riccati Transformation in the Matrix Case," (with P. Nelson and G. M. Wing), Journal of Mathematical Analysis and Applications, **65**, 201-210 (1978).
3. "A Theoretical Result on the Effectiveness of the Addition Formulae for Two-Point Boundary Systems," (with P. Nelson and G. M. Wing), Journal of Mathematical Analysis and Applications, **67**, 392-399 (1979).
4. "A Unique Parameterization for the Transition Potential in Muonic X-Rays," (with S. Wald and Y. N. Kim), Zeitschrift fur Physik, **289A**, 391-394 (1979).
5. "Exact Calculations for Penetrability and the Prediction of the Spontaneous Fission Half-Lives of Superheavy Elements," (with S. Wald and M. A. K. Lodhi), Proceedings of International Conference on Superheavy Elements, Pergamon Press, New York, 467-478, (1979).
6. "Possible Muonic Transitions and X-Ray Spectra from Superheavy Elements," (with S. wald and M. A. K. Lodhi), Proceedings of International Conference on Superheavy Elements, Pergamon Press, New York, 442-450 (1979).
7. "An Automatic Error-Control Technique for Computations of Eigenlengths," (with P. Nelson), Journal of Computational Physics, **37**, 388-398 (1980).
8. "More Corrections to the Sixth Order Anomalous Magnetic Moment of the Muon", (with M. L. Laursen and M. A. Samuel), Zeitschrift fur Physik, **6C**, 3-6 (1980).
9. "Lattice Constant at the Insulator-Metal Transition of Crystalline Xenon," (with S. B. Trickey, R. S. Weidman and A. B. Kunz), Physical Review Letters, **45**, 933-935 (1980).
10. "Oscillation and Comparison Theorems for $y''(x) + \lambda\ a\ (x)\ y(x) = 0$," (with R. M. Anderson), Journal of Mathematical Analysis and Applications, **83**, 100-109 (1981).
11. "Adequacy of Local Exchange Excitation Hamiltonians in Insulators," (with S. B. Trickey and J. P. Worth), Physica Status Solidi, **106**, 613-620 (1981).
12. "Augmented Plane Wave to Gaussian Orbital Conversion Procedure: One Electron States and Compton Profiles of FCC Neon," (with S. B. Trickey), Physical Review B, **24**, 1751-1760 (1981).
13. "BCC-FCC Allotropy, F-Bands, and Metallization in Xenon and Krypton," (with S. B. Trickey and A. B. Kunz), Solid State Communications, **41**, 351-353 (1982).
14. "GCM Treatment of Some Model Periodic Systems," (with L. R. Kauder and S. B. Trickey), International Journal of Quantum Chemistry, **23**, 1355-1361 (1983).
15. "Computation of Electron Transport in Layered Media by Invariant Imbedding," (with P. Nelson and J. C. Garth), Advances in Reactor Computations, American Nuclear Society, **Vol. II**, 623-633 (1983).
16. "Optical Absorption of Solid Xenon at High Pressure," (with S. B. Trickey and A. B. Kunz), Physica Status Solidi, **121**, K47-50 (1984).
17. "Central Depression in ^{4}He Charge Density and Form Factor," (with M. A. K. Lodhi), Proceedings of International Conference on Nuclear Physics, (Tipografia Compositori, Bologna), **Vol. I**, 321 (1984).

18. "The Electronic Structure of Nitromethane: A Multi-Configurational Self-Consistent-Field Study," (with D. S. Marynick, J. L. Fry and D. A. Kleier), Journal of Molecular Structure, **108**, 45-48 (1984).
19. "Computational Study of the Upwind Finite-Difference Method," (with P. Nelson and D. L. Seth), Transactions of the American Nuclear Society, **46**, 436-438 (1984).
20. "A Configuration Interaction Study of Small Lithium Clusters," (with J. L. Fry and C. W. Myles), Journal of Physics B, **18**, 381-386 (1985).
21. "Electronic Structure Calculations on R-NO_2 Species, R=H, CH_3, NH_2, HO and F," (with D. S. Marynick and J. L. Fry), Chemical Physics Letters, **116**, 429-433 (1985).
22. "Surface Effects in Cluster Calculations of Energy Profiles of Mu in Diamond," (with S. Estreicher, J. L. Fry and D. S. Marynick), Physical Review Letters, **55**, 1976-1978 (1985).
23. "A Study of Positron Diffusion and Annihilation in Gases," (with S. R. Kafle, J. L. Fry and S. C. Sharma), Proceedings of Seventh International Conference on Positron Annihilation, World Scientific Publishing, 359-361 (1986).
24. "Minimal Basis-Set SCF-MO Calculations on Aggregates of Lithium and Beryllium Atoms," Journal of Physics B, **19**, 1253-1259 (1986).
25. "Electron Transport in Layered Media by Invariant Imbedding," (with P. Nelson and J. C. Garth), Proceedings of the Conference on Integral Methods in Science and Engineering, Hemisphere Publishing, 37-53 (1986).
26. "Interstitial Hydrogen in Diamond: A Detailed Hartree-Fock Analysis," (with S. Estreicher, J. L. Fry and D. S. Marynick), Physical Review B, **34**, 6071-6079 (1986).
27. "A Hartree-Fock Study of Carbon Clusters," Journal of Physics B, **20**, 5233-5239 (1987).
28. "A Theoretical Study of Carbon Clusters: Equilibrium Geometries and Electronic Structures of C_n" (with L. S. Ott), Proceedings of the NATO Advanced Workshop and International Symposium on the Physics and Chemistry of Small Clusters, Plenum Publishing, New York, 95-101 (1987).
29. "An *Ab Initio* Study of the Interaction of Atomic Hydrogen with Lithium Clusters," (with A. S. Hira and J. L. Fry), Proceedings of the NATO Advanced Workshop and International Symposium on the Physics and Chemistry of Small Clusters, Plenum Publishing, New York, 413-419 (1987).
30. "Response to Comment on Surface Effects in Cluster Calculations of Energy Profiles of Mu in Diamond," (with S. Estreicher, J. L. Fry and D. S. Marynick), Physical Review Letters, **57**, 3301 (1987).
31. "Many Body Perturbation Theory Applied to Hydrogen Interaction with Lithium Clusters," (with A. S. Hira), Physical Review B, **37**, 9943-9950 (1988).
32. "Many Body Perturbation Theory Applied to Small Germanium Clusters," (with M. S. Islam), Chemical Physics Letters, **153**, 496-502 (1988).
33. "A Note on the Bhatia Theory of Liquid NaSn Alloys," (with W. H. Young), Physics and Chemistry of Liquids, **19**, 7-9 (1989).
34. "A Hartree-Fock Study of Ionized Carbon Clusters," (with M. S. Islam), Journal of Physics B, **22**, 2071-2079 (1989).
35. "*Ab Initio* Study of H Chemisorption on Li Clusters in FCC(100) Symmetry," (with A. S. Hira), Physical Review B, **40**, 3507-3513 (1989).
36. "*Ab Initio* Many Body Perturbation Theoretic Study of Small Sodium Clusters," Solid State Communications, **71**, 311-316 (1989).

37. "Short Range Form of the Charge Distribution of ^{4}He and Its Charge Form Factor," (with M. A. K. Lodhi and C. Yalcin), IL Nuovo Cimento, **101A**, 155-162 (1989).
38. "*Ab Initio* Many Body Perturbation Theoretic Study of Sodium Cationic Clusters," Solid State Communications, **72**, 1051-1056 (1989).
39. "*Ab Initio* Many Body Perturbation Theoretic Study of Small Potassium Clusters," (with S. D. Altekar), Physical Review B, **42**, 1444-1447 (1990).
40. "An *Ab Initio* Study of Neutral K_nNa and Cationic K_nNa^+ Clusters," (with S. P. Berry), Journal of Physics B, **23**, 2365-2377 (1990).
41. "An *Ab Initio* Study of Small Potassium Cationic Clusters," (with S. D. Altekar), Physica Status Solidi, **161**, 463-472 (1990).
42. "A Correlation Study of H Interaction with Li Clusters in BCC(111) Symmetry," (with A. S. Hira), Surface Science, **234**, 397-411 (1990).
43. "A Correlation Study of Chemisorption of Atomic Oxygen on Li BCC Clusters," (with A. S. Hira), Journal of Physics B, **24**, 881-896 (1991).
44. "An *Ab Initio* Study of H Chemisorption on Li Clusters in FCC(110) and FCC(111) Symmetries," (with A. S. Hira), Surface Science, **249**, 199-212 (1991).
45. "On the Existence and Stability of Small K_nMg Clusters," (with S. D. Altekar), Journal of Physics: Condensed Matter **3**, 8319-8326 (1991).
46. "A Correlation Study of Large Potassium Cationic Clusters," (with K. M. Kanal and I. A. Howard), Physica Status Solidi, **167**, 465-475 (1991).
47. "Lithium Atom Chemisorption on Silicon Clusters," (with S. N. Ekpenuma), Surface Science, **257**, L659-L664 (1991).
48. "Isomer Shifts and Electric Field Gradients in Y $(Fe_{1-x}Al_x)$," (with P. Blaha and K. Schwarz), Journal of Magnetism and Magnetic Materials, **104**, 683-684 (1992).
49. "Adsorption of Monatomic Hydrogen in Icosahedral Borides," (with I. A. Howard), Physical Review B, **45**, 3680-3684 (1992).
50. "Structure and Binding in Small Neutral and Cationic Boron Clusters," (with I. A. Howard and K. M. Kanal), Physical Review B, **45**, 14247-14255 (1992).
51. "A Correlation Study of Large Potassium Neutral Clusters," (with K. M. Kanal and I. A. Howard), Physica Status Solidi, **171**, 131-140 (1992).
52. "Correlation Study of Sodium Atom Chemisorption on Silicon Surface," (with S. N. Ekpenuma), Journal of Physics: Condensed Matter, **4**, 5077-5086 (1992).
53. "A Correlation Study of Large Neutral and Singly Ionized Carbon Clusters," (with M. S. Islam), Journal of Physics: Condensed Matter, **4**, 4101-4114 (1992).
54. "Stabilities of Adsorption Sites and Charge Transfers at the K/Si (001) Surface," (with S. N. Ekpenuma), Physical Review B, **46**, 16163-16166 (1992).
55. "A Computational Study of the Discretization Error in the Solution of the Spencer-Lewis Equation by Doubling Applied to the Upwind Finite Difference Approximation," (with P. Nelson and D. L. Seth), Journal of Computational Physics, **103**, 370-381 (1992).
56. "A Note on 2-D Electron Transport Using Discrete Ordinates," (with R. P. Datta, A. S. Hira and B. R. Wienke), Supercomputer, **9**, 15-21 (1992).
57. "A Hartree-Fock Study of Small Carbon Cluster Anions," (with B. K. Rao), Physics and Chemistry of Finite Systems: From Clusters to Crystals, Kluwer Academic Publishing, 645-650 (1992).
58. "On the Parallelization of a S_n Transport Algorithm on a Cray Y-MP," (with S. D. Altekar and B. R. Wienke), Parallel Computing, **19**, 823-834 (1993).

59. "On the Interaction of a Sodium Atom with the Silicon (111) Surface," (with M. Mahapatra and A. S. Hira), Journal of Physics: Condensed Matter **5**, 6543-6554 (1993).
60. "A Discrete-Ordinates Study of Two Dimensional Electron Transport," (with R. P. Datta and B. R. Wienke), Journal of Physics D: Applied Physics, **26**, 1077-1083 (1993).
61. "Stability of Clusters in Two Dimensions," (with B. K. Rao and P. Jena), Physical Review B, **42**, 14702-14705 (1993).
62. "A Spatial Characteristic Scheme for Multigroup Discrete Ordinates Electron Transport in Two Dimensions," (with R. P. Datta), Physica Status Solidi, **180**, 85-95 (1993).
63. "Three Particle Correlation Function and Structural Theories of Dense Metallic Liquids," (with K. I. Golden and N. H. March), Molecular Physics, **80**, 915-924 (1993).
64. "Correlation Study of Sodium Atom Chemisorption on the GaAs(110) Surface," (with K. M. Song and D. C. Khan), Physical Review B, **49**, 1818-1826 (1994).
65. "A Diffraction Study of the Structure of Liquid Potassium Near Freezing and Density Functional Theory of Pair Potentials," (with M. W. Johnson, N. H. March and F. Perrot), Philosophical Magazine, **69**, 965-977 (1994).
66. "A Cluster Study of the Stabilities of Adsorption Sites and Charge Transfers at the Na/Si (100) Surface," (with P. C. Lin and A. S. Hira), Physica Status Solidi, **182**, 331-345 (1994).
67. "On the Electronic Structures of GaAs Clusters," (with K. M. Song and P. Khowash), Journal of Physics B, **27**, 1637-1647 (1994).
68. "Exploiting Parallelism in a S_n Transport Algorithm : An Angular Approach." (with S. D. Altekar), Parallel Algorithms and Applications, **4**, 317-328 (1994).
69. "An *Ab Initio* Study of Potassium Chemisorption on GaAs (110) Surface," (with K. M. Song), Journal of Physics: Condensed Matter, **6**, 9571-9583 (1994).
70. "An *Ab Initio* Study of Cesium Chemisorption on the GaAs(110) Surface," (with K. M. Song), Physical Review B, **50**, 14255-14266 (1994).
71. "A Computational Model for Electron Backscattering in Electron Dosimetry," (with T. Kellogg), Medical Physics, **22**, 25-30 (1995).
72. "Geometries and Electronic Structures of Negatively Charged Carbon Clusters," (with B. K. Rao), Zeitschrift fur Physik D, **33**, 197-201 (1995).
73. "Interaction Sites of a Na^+ Ion and a Na Atom with a C_{60} Molecule," (with A. S. Hira), Physical Review A, **52**, 141-148 (1995).
74. "Diatomic Heteronuclear Iron-Metal(3d) Molecule Clusters and their Electronic Structures (with H. J. Choi and A. M. Lemos), Journal of the Korean Physical Society, **28**, 796-798 (1995).
75. "On the Choice of Basis Sets for the Study of Chemisorption of Hydrogen on Aluminum," (with P. C. Lin, J. Niu and B. K. Rao), Proceedings of the International Symposium of the Science and Technology of Atomically Engineered Materials , World Scientific Publishing, 275-279 (1996).
76. "Anomalous Spectroscopy of Li_4^- Cluster," (with B. K. Rao and P. Jena), Physical Review Letters, **76**, 2878-2881(1996).
77. "A Computational Model for Coupled Electron-Photon Transport in Two Dimensions," (with R. P. Datta, S. D. Altekar and J. E. Morel), Physical Review E, **53**, 6514-6522 (1996).
78. "Interactions of the Na_2 Dimer with a C_{60} Molecule," (with A. S. Hira), Physical Review A, **54**, 2205-2215 (1996).

79. "A Cluster Study of Rb Atom Chemisorption on GaAs (110) Surface," (with K. M. Song), Journal of Physics: Condensed Matter, **8**, 6617-6627 (1996).
80. "Geometries and Energies of Small Aluminum Clusters in FCC Symmetry," (with B. K. Rao), Journal of Physics : Condensed Matter, **9**, 2859-2868 (1997).
81. "A Correlation Study of Boron Dimers and Trimers," (with I. A. Howard), Zeitschrift fur Physik D, **42**, 299-301 (1997).
82. "Force Fields in Liquid and Solid Cu Metal: Relation between Quantum Chemical and Density Functional Treatments," (with N. H. March), Physics and Chemistry of Liquids, **36**, 207-214 (1998).
83. "A Density Functional Study of Small Neutral and Cationic Vanadium Clusters V_n and V_n^+ (n=2-9)," (with X. Wu), Journal of Chemical Physics, 110, 2437-2445 (1999).
84. "Si_n Clusters: Surface Energy Considerations for Large n and All-Electron Hartree-Fock Calculations for n=45," (with N. H. March, D. J. Klein and X. Wu), International Journal of Quantum Chemistry, **75**, 829-838 (1999).
85. "A Density Functional Cluster Study of K Adsorption on GaAs (110) Surface," (with M. Panda), Journal of Vacuum Science and Technology A, **17**, 2647-2654 (1999).
86. "An *Ab Initio* Cluster Study of Chemisorption of Atomic Cs on Ga-Rich GaAs (100) (2x1), (2x2) and (4x2) Surfaces," (with R. Schailey), Journal of Chemical Physics, **111**, 8628-8638 (1999).
87. "Possible Existence of the Plutonium Dimer," (with E. Archibong), Physical Review A,**60**, 5105-5107 (1999).
88. "Relation between Vacancy Properties and Surface Energies in Three Noble or Transition Metals (with N. H. March), Journal of Physics and Chemistry of Solids, **61**, 827-828 (2000).
89. "All-Electron Density Functional Theory Calculations of the Zero-Pressure Properties of Plutonium Dioxide," (with J. C. Boettger), AIP Conference Proceedings 532, Plutonium Futures-The Science, edited by K. K. S. Pillay and K. C. Kim , 532-422 (2000).
90. "An *Ab Initio* Study of PuO_2 and PuN_2," (with E. Archibong), Journal of Molecular Structure (TheoChem), **530**, 165-170 (2000).
91. "All Electron LCGTO Calculations for Uranium Dioxide," (with J. C. Boettger), International Journal of Quantum Chemistry, **80,** 824-830 (2000).
92. "A Hybrid Density Functional Cluster Study of the Bulk and Surface Electronic Structures of Plutonium Monoxide," (with X. Wu), Physica B, **293**, 362-375 (2001).
93. "A Hybrid Density Functional Cluster Study of the Bulk and Surface Electronic Structures of Plutonium Dioxide," (with X. Wu), Physica B, **301**, 359-369 (2001).
94. "A Density Functional Study of Plutonium Dioxide," (with X. Wu), European Physical Journal B, **19**, 343-351 (2001).
95. "A Cluster Approach to Hydrogen Chemisorption on GaAs (100) Surface," (with R. Schailey), Computational Materials Science, **22**, 169-179 (2001).
96. "A Density Functional Study of Water Adsorption on PuO_2 (110) Surface," (with X. Wu), Physical Review B, **65**, 085403 (2002).
97. "On the Electronic Structures of Gold Tetramer Clusters," Computational Materials Science, **25**, 279-284 (2002).
98. "An All-Electron LCGTO Study of Square and Hexagonal Plutonium Monolayers," (with J. C. Boettger), European Physical Journal B, 27, 429-433 (2002).

99. "Fully Relativistic Density Functional Calculations on Hydroxylated Actinide Oxide Surfaces," (with J. C. Boettger), International Journal of Quantum Chemistry, **90**, 1470 – 1477 (2002).
100. "A Correlation Study of Small Silver Clusters," (with M. N. Huda), European Physical Journal D, **22**, 217-227 (2003).
101. "Electronic Structures and Magic Numbers of Small Silver Clusters: A Many Body Perturbation Theoretic Study," (with M. N. Huda), Physical Review A, **67,** 013201-1-13 (2003).
102. "Quantum Size Effects in Hexagonal Plutonium Layers," (with J. C. Boettger), AIP Conference Proceedings, **673**, 198-200 (2003).
103. "Carbon Dimer in Silicon Cage: A Class of Highly Stable Silicon Carbide Clusters," (with M. N. Huda), Physical Review A (Rapid Communications), **69**, 011201-4 (2004).
104. "Carbon Dimer in Silicon Cage: A Class of Highly Stable Silicon Carbide Clusters," (with M. N. Huda), Virtual Journal of Nanoscale Science and Technology, Feb. 9 (2004).
105. "A First-Principles Electronic Structure Study of Quantum Size Effects in (111) Films of δ-Plutonium," (with J. C. Boettger), Physical Review B, **70**, 085418-1-6 (2004).
106. "Electronic Structure and Bonding of Oxygen on Plutonium Layers," (with M. N. Huda), European Physical Journal B, **40,** 337-346 (2004).
107. "A Density Functional Study of Atomic Hydrogen Adsorption on Plutonium Layers," (with M. N. Huda), Physica B, **352**, 5-17 (2004).
108. "Novel Silicon-Carbon Fullerene-Like Cages: A Class of sp^2–sp^3 Covalent-Ionic Hybridized Nanosystems", (with M. N. Huda), European Physical Journal D, **31,** 63-68 (2004).
109. "Density Matrix Force-Balance Equation Applied to He, Be, and Ne atoms and to Almost-Spherical Methane-Like Molecules," (with I. A. Howard, N. H. March, F. Bartha, and M.-L. Zhang), International Journal of Quantum Chemistry, **100,** 155-165 (2004).
110. "A Density Functional Study of the Structure and Energetics of Small Hetero-Atomic Silicon-Carbon Nanoclusters," (with P. Pradhan), Journal of Molecular Structure: THEOCHEM, **716**, 109 -130 (2005).
111. "A Density Functional Study of O_2 Adsorption on the (100) Surface of γ - Uranium," (with M. N. Huda), International Journal of Quantum Chemistry, **102,** 98-105 (2005).
112. "A Density Functional Study of Molecular Oxygen Adsorption and Reaction Barrier on Pu (100) Surface," (with M. N. Huda), European Physical Journal B, **43**, 131-141 (2005).
113. "An *Ab Initio* Study on the Use of Ag2- for Catalytic Oxidation of CO," (with P. Dholabhai and X. Wu), Journal of Molecular Structure: THEOCHEM, **723**, 139-145 (2005).
114. "A Cluster Study of Aluminum Adsorption on Ga-Rich GaAs (100) (2x1) and β (4x2) Surfaces," (with M. L. Mayo), European Physical Journal D, **33**, 413-420 (2005).
115. "A Full-Potential-Linearized-Augmented-Plane-Wave Electronic Structure Study of δ - Plutonium and the (001) Surface," (with X. Wu), Physical Review B, **72,** 045115-1-9 (2005).

116. "An *Ab Initio* Study of H_2 Interaction with the Pu (100) Surface," (with M. N. Huda), Physica B, **366,** 95-109 (2005).
117. "Molecular Hydrogen Adsorption and Dissociation on the Pu (111) Surface,' (with M. N. Huda), Physical Review B, **72**, 085101-1-11 (2005).
118. "An *Ab Initio* Study of Molecular Oxygen Adsorption on Pu (111) Surface," (with M. N. Huda), International Journal of Quantum Chemistry, **105,** 280-291 (2005).
119. "Relativistic LCGTO-FF Calculations of Quantum Size Effects in Plutonium Films," (with J. C. Boettger), International Journal of Quantum Chemistry, **105**, 564-570 (2005).
120. "Quantum Size Effects in δ – Pu (110) Films," (with H. R. Gong), European Physical Journal B, **48**, 409-416 (2005).
121. "Silicon-Carbon Fullerene-Like Nanostructures : An *Ab Initio* Study on the Stability of $Si_{60}C_{2n}$ (n = 1, 2) Clusters," (with A. Srinivasan and M. N. Huda), Physical Review A, **72**, 063201-1-10 (2005).
122. "An *Ab Initio* Study of Charge State Effects on the Use of Ag_2^+ and Ag_2 for Catalytic Oxidation of CO," (with P. Dholabhai), Journal of Theoretical and Computational Chemistry, **4,** 1075-1091 (2005).
123. "Novel Silicon-Carbon Fullerene-Like Nanostructures : An *Ab Initio* Study on the Stability of $Si_{54}C_6$ and $Si_{60}C_6$ Clusters," (with A. Srinivasan), Journal of Nanoscience and Nanotechnology, **6**, 43-53 (2006). *Cover page has a copy of one of our figures.*
124. "A Hybrid Density Functional Study of Armchair Si and Ge Nanotubes," (with P. Pradhan), Journal of Computational and Theoretical Nanoscience, **3**, 128-133 (2006).
125. "An *Ab Initio* Study on the Electronic and Geometric Structures of $Si_mC_n^+$ Cationic Nanoclusters," (with P. Pradhan), European Physical Journal D, **37**, 393-407 (2006).
126. "5f Delocalization of Bulk FCC Americium and the (111) Surface : A FP-LAPW Electronic Structure Study," (with D. Gao), Actinides 2005 – Basic Science, Applications and Technology, MRS Symposium Proceedings, **803**, 39-44 (2006).
127. "A First-Principles Study of the (111), (001), and (110) Surfaces of δ – Pu," (with H. R. Gong), Actinides 2005 – Basic Science, Applications and Technology, MRS Symposium Proceedings, **803**, 44-50 (2006).
128. "The 5f Localization – Delocalization in Square and Hexagonal Americium Monolayers – A FP-LAPW Electronic Structure Study," (with D. Gao), European Physical Journal B, **50**, 497-503 (2006).
129. "A Fully-Relativistic Full-Potential-Linearized-Augmented-Plane-Wave Study of the (111) Surface of δ - Pu," (with H. R. Gong), Surface Science, **600**, 2231-2241 (2006).
130. "Silicon-Carbide Nano Clusters: A Pathway to Future Nano – Electronics," (with M. N. Huda), *Review Article,* Journal of Computational and Theoretical Nanoscience, **3**, 315-341 (2006).
131. "A Density Functional Theoretic Study of Novel Silicon-Carbon Fullerene-Like Nanostructures: $Si_{40}C_{20}$, $Si_{60}C_{20}$, $Si_{36}C_{24}$, and $Si_{60}C_{24}$," (with A Srinivasan and M. N. Huda), European Physical Journal D, **39**, 227-236 (2006).
132. "On the Convergence of the Electronic Structure Properties of the FCC Americium (001) Surface," (with D. Gao), Surface Science, **600**, 4941-4952 (2006).
133. "A Full Potential–Linearized Augmented Plane Wave (FP-LAPW) Study of Atomic Carbon, Nitrogen, and Oxygen Chemisorption on the (100) Surface of δ-Pu," (with R. Atta-Fynn), Physica B, **392**, 112 - 126(2007).

134. "A Density Functional Study of Atomic Oxygen and Carbon Adsorptions on (100) Surface of γ – Uranium," (with P. Dholabhai), Physica Scripta, **75**, 506-514 (2007).
135. "Quantum Size Effects in the (0001) Surface of Double Hexagonal Close Packed Americium (with D. Gao), European Physical Journal B, **55,** 13-22 (2007).
136. "An *Ab Initio* Full Potential Fully Relativistic Study of Atomic Carbon, Nitrogen, and Oxygen Chemisorption on the (111) Surface of δ -Pu," (with R. Atta-Fynn), Physical Review B, **75** , 195112-1-13 (2007).
137. "An *Ab Initio* Computational Study of Atomic Adsorption on GaAs Surfaces Modeled by Nanoclusters," (with M. L. Mayo), Journal of Computational and Theoretical Nanoscience, **4,** 443-452 (2007).
138. "Silicon Carbide Nanostructures to Nanotubes," (with M. N. Huda and L. Kleinman), Journal of Computational and Theoretical Nanoscience, **4,** 739-744 (2007); *Cover page has a copy of one of our figures.*
139. "A First-Principles Electronic Structure Study of the High-Symmetry Surfaces of FCC Americium," (with D. Gao), Journal of Alloys and Compounds, **444-445**. 184-190 (2007).
140. "A Density Functional Study of Carbon Monoxide Adsorption on the (100) Surface of γ-Uranium," (with P. Dholabhai), Journal of Alloys and Compounds, **444-445**. 356-362 (2007).
141. "A Fully Relativistic Density Functional Study of the Actinide Nitrides," (with R. Atta-Fynn), Physical Review B, **76**, 115101-1-12 (2007).
142. "Relaxation of the (111) Surface of δ – Pu and Effects on Atomic Adsorption – An *Ab Initio* Study," (with R. Atta-Fynn), Physica B, **400**, 307-316 (2007).
143. "A Hybrid Density Functional Study of Zigzag Silicon Carbide Nanotubes," (with K. Alam), Nanotechnology, **18**, 495706-1-10 (2007).
144. "First Principles Calculations of Localization of *5f* Electrons in Bulk Alpha-Plutonium," (with H. R. Gong), Journal of the Physical Society of Japan, **76**, 114701-1-4 (2007).
145. "A Relativistic Density Functional Study of FCC Americium and the (111) Surface," (with D. Gao), Physical Review B, **77**, 035123-1-9 (2008).
146. "Hybrid Density Functional Study of Armchair Silicon Carbide Nanotubes," (with K. Alam), Physical Review B, **77**, 035436-1-10 (2008).
147. "Hybrid Density Functional Study of Armchair Silicon Carbide Nanotubes," (with K. Alam), Virtual Journal of Nanoscale Science and Technology, February 11 (2008).
148. "Atomic Adsorption on the (020) Surface of α – Pu – A Density Functional Study," (with R. Atta-Fynn), Physical Review B, **77**, 085105-1-11(2008).
149. "A Density Functional Study of Atomic Hydrogen and Oxygen Chemisorption on the Relaxed (0001) Surface of Double Hexagonal Close Packed Americium," (with P. Dholabhai and R. Atta-Fynn), European Physical Journal B, **61**, 261-270 (2008).
150. "A Hybrid Density Functional Study of Zigzag and Chiral Silicon Nanotubes," (with S. Rathi), Journal of Computational and Theoretical Nanoscience, **5**, 464-475 (2008).
151. "Evolution of SiC NanoCluster from Carbon Fullerene : A Density Functional Theoretic Study," (with M. N. Huda), Chemical Physics Letters, **457**, 124-129 (2008).
152. "On the Electronic and Geometric Structures of Armchair GeC Nanotubes – A Hybrid Density Functional Study," (with S. Rathi), Nanotechnology, **19**, 335706 - 335716 (2008).

153. "An *Ab Initio* Study of Molecular Hydrogen Interaction with SiC Nanotube – A Precursor to Hydrogen Storage," (with S. Mukherjee), Journal of Computational and Theoretical Nanoscience, **5**, 1210-1219 (2008).
154. "On the Existence and Stability of Single Walled SiGe Nanotubes," (with S. Rathi), Chemical Physics Letters, **466**, 79-83 (2008)..
155. "An *Ab Initio* Study of the Adsorption and Dissociation of Molecular Oxygen on the (0001) Surface of Double Hexagonal Close Packed Americium," (with P. Dholabhai and R. Atta-Fynn), Physica B, **403**, 4269-4280 (2008).
156. "A First Principles Study of the Interactions of Fe Atom with Single Walled Zigzag SiC Nanotubes," (with K. Alam), Journal of Computational and Theoretical Nanoscience, **6**, 16-29 (2009).
157. "Does Hybrid Density Functional Theory Predict a Non-Magnetic Ground State for δ – Pu?," (with R. Atta-Fynn), Europhysics Letters, **85**, 27008-p1- -p6 (2009).
158. "Adsorption and Dissociation of Molecular Hydrogen on the (0001) Surface of Double Hexagonal Close Packed Americium," (with P. P. Dholabhai), European Physical Journal B, **67**, 183-192 (2009).
159. "A Relativistic DFT Study of Water Adsorption on δ – Pu (111) Surface," Chemical Physics Letters, **470**, 233-239 (2009).
160. "Adsorption and Dissociation of Water on the (0001) Surface of Double Hexagonal Close Packed Americium," (with P. Dholabhai), Physica Status Solidi B, **246**, 1225-1237 (2009).
161. "A First Principles Study of the Adsorption and Dissociation of CO_2 on the δ-Pu (111) Surface," (with R. Atta-Fynn), European Physical Journal B, **70**, 171-184 (2009)..
162. "Interactions of Fe Atom with Single Wall Armchair SiC Nanotubes : An *Ab Initio* Study," (with K. Alam), Journal of Nanoparticle Research, **11,** 1405- 1420 (2009).
163. "Interplay between Spin Polarization, Orbital Polarization, and Spin-Orbit Coupling in Actinides from Pa to Cm," (with F. Islam), Journal of Computational and Theoretical Nanoscience, **6**, 1458-1467 (2009).
164. "Probing the *5f* Electrons in Am-I by Hybrid Density Functional Theory," (with R. Atta-Fynn), Chemical Physics Letters, **482**, 223-227 (2009).
165. "A Computational Study of Endohedral and Exohedral Complexes of Molecular Hydrogen with Single Wall (3,3) to (6,6) Armchair Silicon Carbide Nanotubes," (with S. Mukherjee), Journal of Computational and Theoretical Nanoscience, **7**, 1518-1530 (2010).
166. "A LDA+U Study of the Photoemission Spectra of the Double Hexagonal Close Packed Phases of Am and Cm," (with F. Islam) Solid State Communications, **150**, 938-942 (2010).
167. "Atomic Hydrogen Adsorption on the (020) Surface of α – Pu – A Computational Study," (with F. Islam), Physica Status Solidi B, 1-10 (2010).
168. "On the Magnetic and Thermodynamic Properties of Americium II – A Hybrid Density Functional Theoretic Study," (with J. Wang and Li Ma), Physics Letters A, 374, 4704 – 4712 (2010).
169. "A Hybrid DFT Description of the (0001) Surface of Americium –I (with R. Atta-Fynn), European Physical Journal B, accepted for publication.
170. "Anomalous Dependence of Band Gaps of Binary Nanotubes on Diameters," (with K. Adhikari and M, N, Huda), Journal of Computational and Theoretical Nanoscience, accepted for publication.

171. "A Fully Relativistic Hybrid Density Functional Study of Elemental Actinide Dioxides and Mixed Actinide Dioxides," (with Li Ma and R. Atta-Fynn), Journal of Nuclear Materials, submitted for publication.
172. "On the Existence and Stability of Double-Walled Armchair Silicon Carbide Nanotubes," (with K. Adhikari), Solid State Communications, submitted for publication.
173. "A Hybrid Density Functional Theory Study of $PuO_{2+0.25}$,$UO_{2+0.25}$, $U_{0.5}Pu_{0.5}O_{2+0.25}$," (with Li Ma), European Physical Journal B, submitted for publication.
174. "Applications and Interpretations of Density Functional Theory with Potential U in Atomic and Molecular Systems," (with M. N. Huda), Physical Review A, submitted for publication.
175. "Co-Adsorptions of Hydrogen Molecules with Single Wall Silicon Carbide Nanotubes : An *Ab Initio* Study," (with S. Mukherjee), Journal of Computational and Theoretical Nanoscience, submitted for publication.

RESEARCH PAPERS PRESENTED (CONTRIBUTED AND INVITED)

1. "Radial Shape of the Nuclear Charge Distribution," (with Y. N. Kim), Texas Academy of Sciences, 1977.
2. "Nuclear Short Range Correlation and Photodisintegration of ^{4}He," (with M. A. K. Lodhi), Bulletin of the American Physical Society, **22**, 612 (1977).
3. "Effect of the Short Range Correlation of the Structure of Light Nuclei and Error Resulting from the Approximation," (with M. A. K. Lodhi and Y. W. Yen), Bulletin of the American Physical Society, **23**, 500 (1978).
4. "GCM Treatment of Some Model Periodic Systems," (with L. R. Kauder and S. B. Trickey), Texas Academy of Sciences (1982).
5. "Electron Transport in Layered Media by Invariant Imbedding," (with P. Nelson and J. C. Garth), Bulletin of the American Physical Society, **28**, 428 (1983).
6. "Quantum Size Effects in the Dielectric Function of Small Metallic Particles," (with C. W. Myles), Bulletin of the American Physical Society, **28**, 382 (1983).
7. "Correlation of Exotic Molecules: Using Second Order Many Body Perturbation Theory for Assesment," (with J. T. Waber, A. B. Kunz and R. S. Weidman), Eighth Canadian Symposium on Theoretical Chemistry, Halifax, Nova Scotia, August 6-12 (1983).
8. "Electronic Structure Calculations of Nitrous Acid," (with D. S. Marynick, J. L. Fry and C. J. Liles), Bulletin of the American Physical Society, **29**, 401 (1984).
9. "Positron Diffusion in Molecular Systems," (with S. R. Kafle and J. L. Fry), Bulletin of the American Physical Society, **30,** 404 (1985).
10. "A Comment on the Small Chain Approximation for the Electronic Structure of Polyacetylene," (with D. S. Marynick and J. L. Fry), Bulletin of the American Physical Society, **30**, 524 (1985).
11. "Hartree-Fock Calculation of Electron States in Aggregates of Lithium and Beryllium Atoms," (with J. L. Fry and C. W. Myles), Bulletin of the American Physical Society, **30**, 524 (1985).
12. "Energy Deposition Distribution by Fast Electrons," (with P. Nelson and J. M. Richards), Bulletin of the American Physical Society, **31**, 595 (1986).
13. "Electron Transport in the Quantum Size Effect Regime," (with M. Islam and K. Schwartzman), Bulletin of the American Physical Society, **31**, 595 (1986).

14. "Isotope Shifts in Muonic Atom X-Rays," (with R. M. Marroum), Bulletin of the American Physical Society, **31**, 677 (1986).

15. "Hartree-Fock Calculation of Electron States in Aggregates of Carbon Atoms," Bulletin of the American Physical Society, **31**, 682 (1986).

16. "An *Ab Initio* Study of the Interaction of Atomic Hydrogen with Lithium Clusters," (with A. S. Hira), Poster Presentation, NATO Advanced Workshop and International Symposium on the Physics and Chemistry of Small Clusters, Virginia Commonwealth University, Richmond, Virginia (1986).

17. "A Theoretical Study of Carbon Clusters: Equilibrium Geometries and Electronic Structures of C_n ," (with L. S. Ott), Poster Presentation, NATO Advanced Workshop and International Symposium on the Physics and Chemistry of Small Clusters, Virginia Commonwealth University, Richmond, Virginia (1986).

18. "Monte Carlo Evaluation of Feynman Path Integrals," (with T. E. Cathey), Bulletin of the American Physical Society, **32,** 476 (1987).

19. "Hartree-Fock Calculations of Electron States in Ionized Carbon Clusters," (with M. Islam), Bulletin of the American Physical Society, **32**, 602 (1987).

20. "Chemisorption Studies of the Interactions of a Hydrogen Atom with Lithium Clusters," (with A. S. Hira), Bulletin of the American Physical Society, **32**, 603 (1987).

21. "Hartree-Fock Study of Ionized Carbon Clusters," (with M. Islam), Poster Presentation, Southwest Theoretical Chemistry Conference, Stephenville, Texas (1987).

22. "An *Ab Initio* Study of Li_3, Li_4, and Li_5 Clusters in the FCC(100) Symmetry," (with A. S. Hira), Poster Presentation, Southwest Theoretical Chemistry Conference, Stephenville, Texas (1987).

23. "Ionized Carbon Clusters: An *Ab Initio* Point View," (with M. Islam), Bulletin of the American Physical Society, **33**, 789 (1988).

24. "An *Ab Initio* Study of the Interactions of a Hydrogen Atom with Lithium Clusters in FCC(100) Symmetry," (with A. S. Hira), Bulletin of the American Physical Society, **33**, 816 (1988).

25. "An *Ab Initio* Study of Li_n (n=3-7) Clusters in FCC(100) Symmetry," (with A. S. Hira), Poster Presentation, Fourth International Symposium on Small Particles and Inorganic Clusters, Aix-En-Provence, France (1988).

26. "A Study of the Discretization Error in the Computational Solution of the Spencer-Lewis Equation," Bulletin of the American Physical Society, **35**, 259 (1990).

27. "Many Body Perturbation Theory Applied to Neutral and Cationic Potassium K_n (n=2-7) Clusters," (with S. D. Altekar), Bulletin of the American Physical Society, **35**, 312 (1990).

28. "Chemisorption of Atomic Hydrogen on Lithium Clusters in the BCC (100), (110) and (111) Symmetries," (with A. S. Hira), Bulletin of the American Physical Society, **35**, 312 (1990).

29. "Chemisorption of Atomic Hydrogen on Lithium Clusters in the FCC (100), (110), and (111) Symmetries," (with A. S. Hira), Bulletin of the American Physical Society, **35**, 312 (1990).

30. "*Ab Initio* Studies of Neutral K_nNa, Cationic K_nNa^+, K_nMg, and K_nMg^+ Clusters," (with S. P. Berry and S. D. Altekar), Bulletin of the American Physical Society, **35**, 312 (1990).

31. "An MBPT Study of the Interaction of an O Atom with Lithium Clusters," (with A. S. Hira), Bulletin of the American Physical Society, **35**, 313 (1990).

32. "A Correlation Study of Carbon Clusters," (with M. S. Islam), Bulletin of the American Physical Society, **36**, 353 (1991).

33. "Geometries and Energies of Small Aluminum Clusters in FCC Symmetry," (with B. K. Rao), Bulletin of the American Physical Society, **36**, 506 (1991).

34. "An *Ab Initio* Study of Neutral and Cationic Potassium Clusters," (with K. M. Kanal and I. A. Howard), Bulletin of the American Physical Society, **36**, 653 (1991).

35. "An *Ab Initio* Study of Oxygen Chemisorption on Lithium Clusters," (with A. S. Hira), Bulletin of the American Physical Society, **36**, 854 (1991).

36. "Adsorption of Hydrogen onto Icosohedral Boron Clusters," (with I. A. Howard), Bulletin of the American Physical Society, **36**, 941 (1991).

37. "An *Ab Initio* Study of Alkali Metal Atom Chemisorption on Silicon Clusters," (with S. N. Ekpenuma), Bulletin of the American Physical Society, **36**, 1004 (1991).

38. "A Hartree-Fock Study of Small Carbon Cluster Anions," (with B. K. Rao), Poster Presentation, International Symposium on the Physics and Chemistry of Finite Systems: From Clusters to Crystals, Virginia Commonwealth University, Richmond, Virginia (1991).

39. "Structure and Binding in Small Neutral and Cationic Boron Clusters," (with I. A. Howard and K. M. Kanal), Bulletin of the American Physical Society, **37**, 530 (1992).

40. "A Hartree-Fock Study of Small Carbon Cluster Anions," (with B. K. Rao), Bulletin of the American Physical Society, **37**, 531 (1992).

41. "2-Dimensional Electron Transport Using Discrete Ordinates," (with R. P. Datta and A. S. Hira), Bulletin of the American Physical Society, **37**, 554 (1992).

42. "Sodium Atom Chemisorption on Silicon Surface," (with S. N. Ekpenuma), Poster Presentation, 1992 March Meeting of the American Physical Society.

43. "Structure and Binding in Small Anionic Boron Clusters," (with I. A. Howard), Bulletin of the American Physical Society, **38**, 68 (1993).

44. "Stabilities of Clusters in Two Dimensions," (with B. K. Rao and P. Jena), Bulletin of the American Physical Society, **38**, 69 (1993).

45. "Chemisorption Study of the Interaction of Li and Na Atoms with Si (100) Clusters," (with P. C. Lin and A. S. Hira), Bulletin of the American Physical Society, **38**, 1682 (1993).

46. "An *Ab Initio* Study of Na Chemisorption on the Si (111) Surface," (with M. Mahapatra and A. S. Hira), Bulletin of the American Physical Society, **38**, 1681 (1993).

47. "Spectroscopy of the Li_4^- Cluster," (with B. K. Rao and P. Jena), Bulletin of the American Physical Society, **39**, 415 (1994).

48. "A Correlation Study of Sodium Atom Chemisorption on GaAs (110) Surface," (with K. M. Song and D. C. Khan), Bulletin of the American Physical Society, **39**, 84 (1994).

49. "On the Electronic Structures of GaAs Clusters," (with P. K. Khowash and K. M. Song), Bulletin of the American Physical Society, **39**, 84 (1994).

50. "On the Choice of Basis Sets for the Chemisorption of Hydrogen on Aluminum," (with P. C. Lin, J. Niu and B. K. Rao), International Symposium on the Science and Technology of Atomically Engineered Materials, Virginia Commonwealth University, Richmond, Virginia, 1995.

51. "An *Ab Initio* Study of Alkali Induced Metallization of GaAs (110) Surface," *Invited Lecture*, Fall Meeting of the Texas Section of the American Physical Society, 1996.

52. "A Simulated Annealing Study of Aluminum Clusters," (with C. Gonzales), Bulletin of the American Physical Society, **42**, 462 (1997).
53. "An *Ab Initio* Cluster Study of the Ga Rich GaAs (100) Surface," (with R. Schailey), Bulletin of the American Physical Society, **42**, 785 (1997).
54. "A Density Functional Study of Small Neutral and Cationic Vanadium Clusters," (with X. Wu), Poster Session, 1999 March Meeting of the American Physical Society.
55. "A Study of the Stability and Characterization of Plutonium Dioxide," (with J. C. Boettger), Poster Session, 23rd Annual Actinide Separations Conference, Pasco, Washington, 1999.
56. "An Electronic Structure Study of Solid Plutonium Dioxide," (with J. C. Boettger), Researchers' Conference, Amarillo National Research Center, Amarillo, Texas, 1999.
57. "A Molecular Orbital Study of Plutonium Oxide and Dioxide," (with E. F. Archibong), Researchers' Conference, Amarillo National Research Center, Amarillo, Texas, 1999.
58. "A Hybrid Density Functional Cluster Study of the Bulk and Surface Electronic Structures of Plutonium Monooxide." (with X. Wu), Bulletin of the American Physical Society, **45**, 897 (2000).
59. "An *Ab Initio* Cluster Study of Chemisorption of Atomic Cesium on Ga-Rich GaAs(100) (2x1), (2x2) and β(4x2) Surfaces," (with R. Schailey), Bulletin of the American Physical Society, **45**, 898 (2000).
60. "A Cluster Approach to Hydrogen Chemisorption on the GaAs(100) Surface," (with R. Schailey), Bulletin of the American Physical Society, **45**. 898 (2000).
61. "An Electronic Structure Study of Solid Uranium and Plutonium Dioxides," (with J. C. Boettger), Bulletin of the American Physical Society, **45**, 900 (2000).
62. "An *Ab Initio* Study of Pu – Related Molecular Complexes," (with E. F. Archibong), Bulletin of the American Physical Society, **45**, 960 (2000).
63. "A Density Functional Study of Square and Hexagonal Plutonium Monolayers," (with J. C. Boettger), Bulletin of the American Physical Society, **45**, 1007 (2000).
64. "Density Functional Study of Plutonium Dioxide,"(with X. Wu), Bulletin of the American Physical Society, **46**,766 (2001).
65. "A Hybrid Density Functional Cluster Study of the Bulk and Surface Electronic Structures of PuO_2," (with X. Wu), Bulletin of the American Physical Society, **46**, 486 (2001).
66. "Density Functional Calculations on Hydroxylated Actinide-Oxide Surfaces," (with J. C. Boettger), Sanibel Symposium (2002).
67. "An Electronic Structure Study of the (110) Surface of δ - Pu," (with M. Leonard), Bulletin of the American Physical Society, 47, 368 (2002).
68. "A Correlation Study of Small Silver Clusters," (with M. N. Huda), Bulletin of the American Physical Society, **47,** 1249 (2002).
69. "*Ab Initio* Modeling of the Endohedral and Exohedral C_{60} Na_2^+ Complexes," (with A. S. Hira), Bulletin of the American Physical Society **47**, 1196 (2002).
70. "Anisotropy of Quantum Size Effects in Aluminum (100), (110), (111) Surfaces", (with M.N.Huda), Bulletin of the American Physical Society, **48**, 1029 (2003).
71. "Quantum Size Effects in Thorium (100) and (111) Surfaces " (with M. N .Huda), Bulletin of the American Physical Society, **48**, 1029 (2003).
72. "An *Ab Initio* Study of Alkali-C_{60} Complexes," (with N. E. Frick and A. S. Hira), Bulletin of the American Physical Society **48**, 106 (2003).

73. "Electronic Structures and Magic Numbers of Small Silver Clusters: A Many Body Perturbation Theoretic Study" (with M. N. Huda), Bulletin of the American Physical Society **48**, 324 (2003).

74. "An *Ab Initio* Cluster Study of Chemisorption of Atomic Oxygen on Ga-Rich GaAs (100) (2x1) and β (4x2) Surfaces," (with M. L. Mayo), Bulletin of the American Physical Society **48**, 489 (2003).

75. "Quantum Size Effects in Hexagonal Plutonium Layers," (with J. C. Boettger), Poster Presentation, Pu2003 Conference, Albuquerque, New Mexico, 2003.

76. "Carbon Dimer in Silicon Cage: A Class of Highly Stable Silicon Carbide Clusters," (with M. N. Huda), Bulletin of the American Physical Society, **49**, 381 (2004).

77. "Novel Silicon-Carbon Fullerene Like Cages: A Class of sp^2-sp^3 Covalent-Ionic Hybridized Nanosystems," (with M. N. Huda), Bulletin of the American Physical Society, **49**, 384 (2004).

78. "Electronic and Bonding Properties of Hydrogen and Oxygen Adsorption on Uranium Layers," (with M. N. Huda), Bulletin of the American Physical Society, **49**, 949 (2004).

79. "Adsorption of Atomic Hydrogen and Oxygen on δ - Pu Layers," (with M. N. Huda), Bulletin of the American Physical Society, **49**, 950 (2004).

80. "An *Ab Initio* Study of Aluminum Chemisorption on Ga-Rich GaAs (100) (2x1) and β (4x2) Surfaces," (with M. L. Mayo), Bulletin of the American Physical Society **49**, 1095 (2004).

81. "Structure and Bonding in Mixed Silicon-Carbon Nanoclusters: A LDA-DFT Study," (with P. Pradhan), Bulletin of the American Physical Society **49**, 1135 (2004).

82. "An *Ab Initio* Study of H_2 Interaction with the Pu (100) Surface," (with M. N. Huda), Bulletin of the American Physical Society **50**, 378 (2005).

83. "An LAPW Study of δ - Plutonium and the (001) Surface," (with X. Wu), Bulletin of the American Physical Society, **50**, 472 (2005).

84. "A First-Principles Electronic Structure Study of Quantum Size Effects in (111) Films of FCC Plutonium," (with J. C. Boettger), Bulletin of the American Physical Society, **50**, 472 (2005).

85. "An *Ab Initio* Study of Molecular Oxygen Adsorption on Pu (111) Surface," (with M. N. Huda), Bulletin of the American Physical Society, **50**, 472 (2005).

86. "In – Rich Reconstructions of the InSb (100) Surface and Chemisorption of Lithium on the c (8x2) Surface – An *Ab Initio* Study ," (with L. G. Sridevi), Bulletin of the American Physical Society, **50**, 140 (2005).

87. "Novel Silicon-Carbon Nanostructures – Electronic Structure Study on the Stability of $Si_{60}C_{2n}$ Clusters," (with A. Srinivasan and M. N. Huda), Bulletin of the American Physical Society, **50**, 302 (2005).

88. "Electronic Structure Studies of Silicon-Carbon Cationic Nanoclusters," (with P. Pradhan), Bulletin of the American Physical Society, **50**, 1169 (2005).

89. "A Density Functional Study of Molecular Oxygen Adsorption and Reaction Barrier on Pu (100) Surface," (with M. N. Huda), Bulletin of the American Physical Society, **50**, 1271 (2005).

90. "Hydrogen Molecule Adsorption and Dissociation on Pu (111) Surface," (with M. N. Huda), Bulletin of the American Physical Society, **50**, 1257 (2005).

91. "An *Ab Initio* Study of the Use of Ag_2^- for Catalytic Oxidation of CO," (with P. Dholabhai), Bulletin of the American Physical Society, **50**, 1444 (2005).
92. "An *Ab Initio* Study of Silicon and Germanium Nanotubes," (with P. Pradhan), Bulletin of the American Physical Society, **50**, 1288 (2005).
93. "Fully Relativistic LCGTO-FF Calculations of Quantum Size Effects in Plutonium Films," (with J. C. Boettger), paper presented at the Sanibel, 2005 Meeting.
94. "A Fully-Relativistic Density Functional Study of the Role of 5f Electrons in Chemical Bonding in Transuranium Elements," *Invited* Oral Presentation at the Contractors' Meeting of the Separations and Heavy Element Chemistry Programs, Division of Chemical Sciences, Geosciences, and Biosciences, United States Department of Energy, Rockville, Maryland (2005).
95. "Novel Silicon-Carbon Nanostructures – An *Ab Initio* Study on the Stability of $Si_{60}C_{20}$ Clusters," (with A. Srinivasan and M. N. Huda), Poster Session, Fall 2005 Meeting of the Materials Research Society, Boston, MA.
96. "5f Delocalization of Bulk FCC Americium and the (111) Surface: FP-LAPW Electronic Structure Calculations," (with D. Gao), Oral Presentation, Fall 2005 Meeting of the Materials Research Society, Boston, MA.
97. "A First-Principles Study of the (111), (001), and (110) Surfaces of Delta-Pu," (with H. Gong), Oral Presentation, Fall 2005 Meeting of the Materials Research Society, Boston, MA.
98. "An *Ab Initio* Study of Atomic Oxygen Adsorption on the (100) Surface of Gamma-Uranium," (with P. Dholabhai), Poster Session, Fall 2005 Meeting of the Materials Research Society, Boston, MA.
99. "A Density Functional Study of the Americium (001) Surface," (with D. Gao), Bulletin of the American Physical Society, **51**, 480 (2006).
100. "A Density Functional Study of Carbon Monoxide Adsorption on the (100) Surface of γ – Uranium," (with P. Dholabhai), Bulletin of the American Physical Society, **51**, 480 (2006).
101. "A Density Functional Study of Atomic Oxygen and Carbon Adsorptions on the (100) Surface of γ – Uranium," (with P. Dholabhai), Bulletin of the American Physical Society, **51**, 480 (2006).
102. "Quantum Size Effects in δ – Pu (111) and (110) Films," (with H. R. Gong), Bulletin of the American Physical Society, **51**, 168 (2006).
103. "Novel Silicon-Carbon Nanostructures: A DFT Study on the Stability of $Si_{60}C_{2n}$ (n=3, 10, 12) Cluster," (with A. Srinivasan and M. N. Huda), Bulletin of the American Physical Society, **51**, 116 (2006).
104. "Silicon-Carbon Nanostructures to Nanotubes," (with M. N. Huda and L. Kleinman), Bulletin of the American Physical Society, **51**, 928 (2006).
105. "The Ground State and Localization of 5f Electrons in α – Pu," (with H. R. Gong), Poster Presentation, Pu Futures – The Science, July 9-13, 2006, Pacific Grove, California.
106. "A First-Principles Electronic Structure Study of the High-Symmetry Surfaces of FCC Americium," (with D. Gao), *Invited Talk*, Pu Futures – The Science, July 9-13, 2006, Pacific Grove, California.
107. "A Density Functional Study of Carbon Monoxide Adsorption on the (100) Surface of Gamma-Uranium," (with P. Dholabhai), Poster Presentation, Pu Futures – The Science, July 9-13, 2006, Pacific Grove, California.

108. "A Comparative *Ab Initio* Electronic Structure Study of FCC δ – Pu and Am II Surfaces," *Invited* Poster Presentation, Third United States Department of Energy – Russian Academy of Sciences Workshop on Basic Actinide Chemistry, July 13-15, 2006, Monterey, California.

109. "A Hybrid Density Functional Study of Armchair SiC Nanotubes," (with K. Alam), Paper Presented at the Fall, 2006 Joint Meeting of the Texas Section of the American Physical Society, October 5-7, 2006, Arlington, Texas.

110. "Relaxation of Actinide Surfaces: An All Electron Study," (with R. Atta-Fynn and P. Dholabhai), Paper Presented at the Fall, 2006 Joint Meeting of the Texas Section of the American Physical Society, October 5-7, 2006, Arlington, Texas.

111. "An *Ab Initio* Full Potential Fully Relativistic Study of Atomic Carbon, Nitrogen, and Oxygen Chemisorption on the (111) Surface of δ – Pu," (with R. Atta-Fynn), Paper Presented at the Fall, 2006 Joint Meeting of the Texas Section of the American Physical Society, October 5-7, 2006, Arlington, Texas.

112. "A FP-LAPW Study of Atomic Carbon, Nitrogen, and Oxygen Chemisorption on the (100) Surface of δ – Pu," (with R. Atta-Fynn), Paper Presented at the Fall, 2006 Joint Meeting of the Texas Section of the American Physical Society, October 5-7, 2006, Arlington, Texas.

113. "A Hybrid Density Functional Study of Si Nanotubes," (with S. Rathi), paper presented at the March, 2007 Meeting of the American Physical Society.

114. **"**A Hybrid Density Functional Study of SiC Nanotubes,**"** (with K. Alam), paper presented at the March, 2007 Meeting of the American Physical Society.

115. "A FP-LAPW Study of Atomic Chemisorption on the (111) Surface of δ - Pu," (with R. Atta-Fynn), paper presented at the March, 2007 Meeting of the American Physical Society.

116. "A FP-LAPW Study of Atomic Chemisorption on the (100) Surface of δ - Pu," (with R. Atta-Fynn), paper presented at the March, 2007 Meeting of the American Physical Society.

117. "An *Ab Initio* Full Potential Fully Relativistic Study of the (0001) Surface of Double Hexagonal Close Packed Americium," (with D. Gao), paper presented at the March, 2007 Meeting of the American Physical Society.

118. "A Fully-Relativistic Density Functional Study of the Role of 5f Electrons in Chemical Bonding in Transuranium Elements," *Invited* Poster Presentation at the Contractors' Meeting of the Separations and Heavy Element Chemistry Programs, Division of Chemical Sciences, Geosciences, and Biosciences, United States Department of Energy, Annapolis, Maryland (2007).

119. "A Fully-Relativistic Density Functional Study of the Role of 5f Electrons in Chemical Bonding in Transuranium Elements," *Invited Speaker,* Contractors' Meeting of the Condensed Phase and Interfacial Molecular Sciences Programs, Division of Chemical Sciences, Geosciences, and Biosciences, United States Department of Energy, Warrenton, Virginia (2007).

120. **"**A Hybrid Density Functional Study of Zigzag SiC Nanotubes." (with K. Alam), poster presented at the March, 2008 Meeting of the American Physical Society, http://meetings.aps.org/Meeting/Mar08, p. 262.

121. "A Density Functional Study of Atomic Hydrogen and Oxygen Chemisorptions on the (0001) Surface of Double Hexagonal Close Packed Americium," (with P. Dholabhai and

R. Atta-Fynn), paper presented at the March, 2008 Meeting of the American Physical Society, http://meetings.aps.org/Meeting/Mar08, p. 435.

122. "Adsorption and Dissociation of Molecular Oxygen on the (0001) Surface of Double Hexagonal Close Packed Americium," (with P. Dholabhai and R. Atta-Fynn), poster presented at the March, 2008 Meeting of the American Physical Society, http://meetings.aps.org/Meeting/Mar08, p. 142.

123. "On the Electronic Structures of Ge-Based Nanotubes," (with S. Rathi), poster presented at the March, 2008 Meeting of the American Physical Society, http://meetings.aps.org/Meeting/Mar08, p. 136.

124. "Atomic Adsorptions on the (020) Surface of α – Pu: A Density Functional Study," (with R. Atta-Fynn), poster presented at the March, 2008 Meeting of the American Physical Society, http://meetings.aps.org/Meeting/Mar08, p. 144.

125. "Relaxation of the (111) Surface of δ –Pu and Effects on Atomic Adsorption: An *Ab Initio* Study," (with R. Atta-Fynn), paper presented at the March, 2008 Meeting of the American Physical Society, http://meetings.aps.org/Meeting/Mar08, p. 196.

126. "On the Electronic and Geometric Structures of Armchair GeC Nanotubes: A Hybrid Density Functional Study," (with S. Rathi). paper presented at the March, 2008 Meeting of the American Physical Society, http://meetings.aps.org/Meeting/Mar08, p. 358.

127. "An *Ab Initio* Study of the Interaction of Transition Metal Atoms with Single-Wall Armchair SiC Nanotubes," (with K. Alam), paper presented at the March, 2008 Meeting of the American Physical Society, http://meetings.aps.org/Meeting/Mar08, p. 514.

128. "Evolution of SiC Nano-Cluster from Carbon Fullerene: A Density Functional Theory Study." (with M. N. Huda), poster presented at the March, 2008 Meeting of the American Physical Society, http://meetings.aps.org/Meeting/Mar08.

129. "A Hybrid Density Functional Study of Pure Ge and GeC Nanotubes: Zigzag Configuration," (with S. Rathi), poster presented at the March, 2009 Meeting of the American Physical Society, http://meetings.aps.org/Meeting/MAR09, p144.

130. "Adsorption and Dissociation of Water on the (0001) Surface of DHCP Americium," (with P. Dholabhai), poster presented at the March, 2009 Meeting of the American Physical Society, http://meetings.aps.org/Meeting/MAR09, p. 277.

131. "Hydrogen Adsorption on the (020) Surface of alpha-Pu : A Computational Study," (with M. Islam), poster presented at the March, 2009 Meeting of the American Physical Society, http://meetings.aps.org/Meeting/MAR09, p 277

132. "A First Principles Study of the Adsorption and Dissociation of CO2 on the delta-Pu (111) Surface," (with R. Atta-Fynn), poster presented at the March, 2009 Meeting of the American Physical Society, http://meetings.aps.org/Meeting/MAR09, p 278.

133. "A Hybrid Density Functional Study of delta-Pu," (with R. Atta-Fynn), paper presented at the March, 2009 Meeting of the American Physical Society, http://meetings.aps.org/Meeting/MAR09, p 313

134. "An LDA+U Study of the Photoemission Spectra of the Ground State Phases of Americium and Curium, "(with M. Islam), paper presented at the March, 2009 Meeting of the American Physical Society, http://meetings.aps.org/Meeting/MAR09, p 278.

135. "Does Water Adsorb Molecularly or Dissociatively on a Plutonium Surface," (with R. Atta-Fynn), paper presented at the March, 2009 Meeting of the American Physical Society, http://meetings.aps.org/Meeting/MAR09, p 364.

136. "On the Interplay between Spin Polarization, Orbital Polarization, and Spin-Orbit Coupling in Actinides from Pa to Cm," (with M. Islam), poster presented at the March, 2009 Meeting of the American Physical Society, http://meetings.aps.org /Meeting/MAR09, p 399.

137. "Adsorption and Dissociation of Molecular Hydrogen on the (0001) Surface of DHCP Americium," (with P. Dholabhai), paper presented at the March, 2009 Meeting of the American Physical Society, http://meetings.aps.org/Meeting/MAR09, p 417.

138. "Can Silicon Carbide Nanotubes be Effective Medium for Hydrogen Storage," (with S. Mukherjee), paper presented at the March, 2009 Meeting of the American Physical Society, http://meetings.aps.org/Meeting/MAR09, p 505.

139. "A Fully-Relativistic Density Functional Study of the Role of 5f Electrons in Chemical Bonding in Transuranium Elements," *Invited* Oral Presentation at the Contractors' Meeting of the Separations and Heavy Element Chemistry Programs, Division of Chemical Sciences, Geosciences, and Biosciences, United States Department of Energy, Warrenton, Virginia (2009).

140. "Electronic, Structural, and Thermodynamic Properties of Mixed Actinide Dioxides (U, Pu, Am) O_2 from Hybrid Density Functional Theory," (with L. Ma), poster presented at the March, 2010 Meeting of the American Physical Society, http://meetings.aps.org/Meeting/MAR10, Bulletin of the American Physical Society, **65,** p 161.

141. "A Hybrid DFT Description of the Ground State Properties of Americium I," (with R. Atta-Fynn), poster presented at the March, 2010 Meeting of the American Physical Society, http://meetings.aps.org/Meeting/MAR10, Bulletin of the American Physical Society, **65,** p 161.

142. "A Hybrid Density Functional Study of Double-Walled Armchair SiC Nanotubes," (with K. Adhikari), paper presented at the March, 2010 Meeting of the American Physical Society, http://meetings.aps.org/Meeting/MAR10, Bulletin of the American Physical Society, **65,** p 350.

143. "Electronic, Structural, and Thermodynamic Properties of Actinide Dioxides, "(with L. Ma and R. Atta-Fynn), paper presented at the March, 2010 Meeting of the American Physical Society, http://meetings.aps.org/Meeting/MAR10, Bulletin of the American Physical Society, **65,** p 443

144. "Quantum Size Effects in the Properties of Non-Magnetic Americium-I (0001) Surface: A Hybrid DFT Study, " (with R. Atta-Fynn), paper presented at the March, 2010 Meeting of the American Physical Society, http://meetings.aps.org/Meeting/MAR10, Bulletin of the American Physical Society, **65,** p 580.

145. "*Ab Initio* Studies of Atomic and Molecular Adsorptions on Plutonium Surfaces," Plenary Speaker, Pu Futures 2010 Conference, Keystone, Colorado, 20-23 September, 2010. Abstract published in p. 212-213, American Nuclear Society, ISBN 978-0-89448-082-9.

SEMINARS AND INVITED PRESENTATIONS:

Brooklyn College of the City University of New York

Condensed Phase and Interfacial Molecular Sciences Meeting, Department of Energy

Department of Energy Laboratory, Idaho Falls

Indian Association for the Cultivation of Science, Calcutta, India

Mathematics Department, University of Texas at Arlington

Michigan Technological University

Naval Research Laboratory, Washington, D. C.

Physics Department, University of Texas at Arlington

Princeton University

Saha Institute of Nuclear Physics, Calcutta, India

Tarleton State University

Technical University of Vienna, Austria

Texas Tech University

BOOKS AUTHORED

"*Probing 5f Electrons: A Relativistic DFT Study of Americium Surfaces,*" (with P. Dholabhai), Lambert Academic Publishing, Germany, 2009.

BOOKS EDITED

"*Nanoclusters and Nanostructured Surfaces,*" (to be published in 2010 by American Scientific Publishers).

EDITORIAL SERVICE

Associate Editor, *Journal of Nanoscience and Nanotechnology*, American Scientific Publishers.
Associate Editor, *Journal of Computational and Theoretical Nanoscience,* American Scientific Publishers.
Member of the Editorial Board, *Transport Theory and Statistical Physics*, Taylor and Francis Publishing.
Member of the Editorial Advisory Board, *Recent Patents on Materials Science*, Bentham Science Publishers.

Member of the Editorial Board, *Journal of Nano Energy and Power Research*, American Scientific Publishers.

REFEREE SERVICE

Canadian Journal of Physics
Chemical Physics Letters
European Physical Journal B
Heteroatom Chemistry
IEEE – Transactions on Electron Devices
International Journal of Quantum Chemistry
Israel Science Foundation
Journal of Alloys and Compounds
Journal of Applied Physics
Journal of Nanoscience and Nanotechnology
Journal of Computational Chemistry
Journal of Computational and Theoretical Nanoscience
Journal of Molecular Structure: Theochem
Journal of Nuclear Materials
Journal of Photochemistry and Photobiology
Journal of Physics A
Journal of Physics B
Journal of Physical Chemistry
Journal of Physics: Condensed Matter
Journal of the Electrochemical Society
National Science Foundation
National Science Foundation Review Panel
Optics Communications
Petroleum Research Foundation
Physica B
Physical Review B
Physical Review Letters
Physics Letters A
Proceedings of the International Symposium on the Physics and Chemistry of Small Clusters, Richmond, Virginia
Proceedings of the Materials Research Society
Surface Science
Session Chair, Non-Metallic Clusters, March 1992 Meeting of the American Physical Society
Session Chair, Fall 2006 Joint Meeting of the Texas Section of the American Physical Society
Texas Journal of Science
Transport Theory and Statistical Physics
United States Department of Energy

United States Department of Energy, Innovative and Novel Computational Impact on Theory and Experiment (INCITE) Scientific Review Panel, 2007.
United States Civilian Research and Development Foundation

Grants and Related Activities

"Safety Analysis for Pipeline Transmission of Hydrogen," Los Alamos Scientific Laboratory, $5,000.00 (1978-79).

Shell Faculty Grant, Michigan Technological University, $250.00 (1981-82).

"Total Energy of Solids and the Generator Coordinate Method", Michigan Technological University Summer Faculty Grant, $2,500.00 (1982).

"Properties of Three-Dimensional Energetic Solids and Molecular Crystals," (with A. B. Kunz, D. R. Beck, R. S. Weidman and T. O. Woodruff), Office of Naval Research, $60,000.00 (1981-82).

"Computation of Electron and Photon Transport in Layered Media by Invariant Imbedding," (with P. Nelson), Air Force Office of Scientific Research, $36,000.00 (1983-84).

"Theoretical Studies of Solids Under Extreme Conditions," (with J. L. Fry, P. C. Pattnaik and G. L. Fletcher), Air Force Office of Scientific Research, $247,000.00 (1983-85).

"Transport Properties of $(Bi,Sb)_2$ $(Te,Se)_3$," (with J. L. Fry and G. L. Fletcher), Marlow Industries, $30,000.00 (1984).

Organized Research Fund, The University of Texas at Arlington, $2,000.00 (1984).

Summer Faculty Grant, Organized Research Fund, The University of Texas at Arlington, $5,000.00 (1985).

Research Travel Grant, The University of Texas at Arlington, $1,000.00 (1986).

Cray Research Corporation Colloquium Grant (with A. H. Weiss), $2,000.00 (1986-87).

"Analysis of Eruption Clouds Detected by Satellites," (with S. Self), National Science Foundation, $78,000.00 (1987-88).

"Molecules to Clusters to Solids – An *Ab Initio* Study," The Welch Foundation, $247,000.00 (1987-96).

Research Travel Grant, The University of Texas at Arlington, $500.00 (1987).

Research Travel Fellowship, University of East Anglia, Norwich, United Kingdom, \$2,000.00 (1987).

Research Travel Grant, The University of Texas at Arlington, \$500.00 (1988).

Research Initiation Grant for 20 CPU hours, The John von Neumann National Supercomputer Center (1988).

"An Analysis of Cluster Physics," Research Support Grant, 500 CPU Hours on an IBM 3090 and \$5,000.00 Travel Expenses, IBM Los Angeles Scientific Center (1988).

Cray Research Corporation Colloquium Grant (with A. H. Weiss), \$5,000.00 (1987-89).

"Computational Techniques for Microscopic Models of Transport in Semiconductors", Cray Research, \$66,000.00 (1989-90).

Research Travel Grant, The University of Texas at Arlington, \$1,500.00 (1990).

"A New Approach to Transport Theoretic Analysis of Radiation Damage in Microelectronic Structures," (with P. Nelson), Texas Advanced Technology Program, \$121,819.00 (1990-91).

Guest Professorship, The Technical University of Vienna, Austria, \$5,000.00 (1990).

Visiting Research Fellowship, Oxford University, England, \$2,000.00 (1991).

"A Computational Database for Scattering Cross-Sections in Semiconductors," 325 CPU Hours on a Cray YMP 8/64, Cray Research (1992).

"A Parallel Transport Algorithm for the Cray YMP," 75 CPU hours on a Cray YMP 8/64, Cray Research (1992).

"Molecules to Clusters to Solids- An *Ab Initio* Study," National Science Foundation, Pittsburgh Supercomputing Center, 600 CPU Hours on a Cray C90 (1994).

"A Theoretical Study of Alkali Atom Interactions with Silicon Surfaces," Cray Research, \$7,500.00 Direct Funds and \$30,000.00 in Computer Time (1995).

"A Density Functional Study of Alkali Interactions with the GaAs (110) Surface," 300 CPU Hours on a Cray C90, National Science Foundation Pittsburgh Supercomputing Center (1995).
"Parallel Electron Transport Softwares on a Cray T3D, 10,000 Units on a Cray T3D, National Science Foundation Pittsburgh Supercomputing Center (1995).

"Acquisition of Quantum: A High Performance SGI Workstation," (with J. L. Fry and I. A. Howard), $50,000.00, Office of Computing and Information Technology, The University of Texas at Arlington (1995).

"An *Ab Initio* Study of Chemisorption on Metal and Semiconductor Surfaces," 5000 units on a SGI Power Challenge and 1000 units on a HP Convex Exemplar, National Center for Supercomputing Applications, Urbana, Illinois (1996).

"Acquisition of Personal Computers," $30,000.00, Office of Computing and Information Technology, University of Texas at Arlington (1996).

"Numerical Discretization Schemes for the Fokker-Planck Equation," Los Alamos National Laboratory, Department of Energy, $24,576.00 (1997).

Visiting Research Fellowships, Princeton University, Princeton, New Jersey, $12,000.00 (1998-99).

"Modeling of Solid State and Thermodynamic Properties of Actinides," (with J. M. Sanchez), $151,864.00, Amarillo National Research Center, Department of Energy (1999).

"A Density Functional Study of the Surface Electronic Behavior of Actinide Metals," $480,000.00 (2001-2010), Welch Foundation, Houston, Texas.

"A Fully Relativistic Density Functional Study of the Role of 5f Electrons in Chemical Bonding in Transuranium Elements," $694,000.00 (2003-2010), United States Department of Energy.

PROPOSALS WRITTEN (NOT FUNDED)

"Theoretical Studies of Energetic Materials," (with D. S. Marynick and J. L. Fry).

"Novel Theoretical Approaches to the Studies of Conducting Polymers." (with D. S. Marynick and J. L. Fry).

"Studies of Conducting Polymers".

"Modeling of Ultra Small Semiconductor Devices," (with P. Nelson, C. W. Myles and H. D. Victory).

"Physics of Sub-Micron Semiconductor Devices and Hot Electron Transport," (with C. W. Myles).

"Ab Initio Studies of Metal Clusters".

"Integration of Code Systems AMBER, CHARM and GAMESS".

"Positron Studies of Defects in Compound Semiconductors," (with R. N. West and J. H. Kaiser).

"A New Approach to the Study of Chemisorption on Metal Surfaces".

"A Computational Database for Scattering Cross-Sections in Semiconductors".

"A Parallel Transport Algorithm for the Cray YMP".

"Design of New Materials from Studies of Interactions of Heteroatoms with Conventional and Novel Semiconductors," (with I. A. Howard).

"Parallel Programming in Electron Transport: Experiments in SIMD and MIMD Architectures".

"Damage Prevention, Detection, and Growth Retardation in Smart Laminated Composites," (with S. P. Joshi and W. S. Chan).

"Impurities in Borides and Small Boron Clusters," (with I. A. Howard).

"An Ab Initio Study of Induced Metallization at Semiconductor Surfaces".

"Serial and Parallel Programming for Electron Transport in Multi-Dimensions".

"An Ab Initio Study of Atomic and Molecular Chemisorption on Aluminum Surfaces".

"Transition Metal Clusters: A Parallel Monte Carlo Simulated Annealing Study".

"Acquisition of Advanced Computer Systems for Visualization of Complex Physical Phenomena," (with T. Black, R. Magnusson, T. Maldonado and Q. Zhang).

"An Ab Initio Study of Metal Oxides and Molecular Adsorption"

"A Novel Approach to Electron and Photon Transport in Biological Tissues," (with R. Glickman, R. Horomoto and D. Sardar).

"First-Principles Investigations on the Deposition and Growth of Novel Dielectric Thin-Film Materials," (with Q. Zhang).

"Novel Parallel Algorithms for Nuclear Well Logging: The PANTHER Consortium," (with P. Nelson and N. Abdurrahman).

"Acquisition of Advanced Computing Facilities for Investigations of Complex Condensed Phase Systems," (with I. A. Howard and Q. Zhang).

"Novel Techniques for the Simulation of Particle Transport on Modern Parallel Computational Platforms," (with P. Nelson, S. Oliviera and N. Abdurrahman).

"Development of Parallel Algorithms for Three-Dimensional Transport Codes," (with S. Oliviera and N. Abdurrahman).

"Coupled Electron-Photon Transport in Multi-Dimensions: A Parallel Discrete Ordinates Study".

"A Novel Cluster Study of Oxide Surfaces and Molecular Adsorption".

"Novel CBN- and AlSiP- Based Materials," (with D. J. Klein).

"Boron Diffusion in Silicon-Based Novel Dielectric Materials," (with Q. Zhang).

"First-Principles Investigation of Novel Dielectric Materials," (with Q. Zhang).

"Towards A Systematic Understanding of the Electronic Structures of the Actinide Molecular Complexes"

"Development of Tools for Nanoparticle Based Technologies"

"Quasi-Continuum Modeling Limits for Nanoscale Systems"

"Mechanisms of Nanoscale Breakdowns in the Vicinity of a Macroscopically Defined Interface Crack," (with S. P. Joshi)

"Structure Property Relations in Nanostructured Alumina-Titania Composites," (with S. P. Joshi)

"Physics Analysis Frontier Center," (with A. White, K. De, and A. Brandt)

"A Novel Algorithm for Direct 4-Component Relativistic Self-Consistent-Field (SCF) Method : Applications to Condensed Matter"

"Acquisitions of Advanced Computing Facilities for Investigations of Complex Condensed Phase Systems"

"Simulations in Electronic Structures for Silicon-Carbide sp^2-sp^3 Hybridized Nanosystems"
"A Novel Class of Carbon Embedded Silicon Nanostructures"

"Integrative Graduate Education and Research in Super- and Parallel- Computing in Particles and Fluids," (with Q. Zhang, Y. Su, C. Liu, H. Shan, J. Zhu, and J. Horwitz).

"Beyond Carbon Nanotubes: A Hybrid Density Functional Electronic Structure Study," (with L. Burggraf)

"Towards the Next Generation of Scintillators: An Ab Initio Study"

"An Ab Initio Full Potential Fully Relativistic Electronic Structure Study of Actinide Nitrides as Nuclear Fuels"

"Beyond Standard Density Functional Theory: Hybrid Density Functional Theory and LDA+U Methods for Applications to Actinides and Actinide Compounds"

"Towards a Complete understanding of the Thermodynamic Properties of UO2, PuO2, and MOX Advanced Actinide Fuels based on Ab Initio Quantum Mechanical Studies"

Post-Doctoral Fellows Directed

Dr. E. Archibong
Dr. R. Atta-Fynn
Dr. A. Bakshai
Dr. R. P. Datta
Dr. S. N. Ekpenuma
Dr. D. Gao
Dr. H. R. Gong
Dr. A. S. Hira
Dr. S. R. Kafle
Dr. D. C. Khan
Dr. M. Mahapatra
Dr. J. Wang
Dr. X. Wu

Graduate Students Directed	Thesis/Dissertation Titles
Mr. T. Cathey (M. S. 1987)	"Numerical Evaluation of Feynman Propagators in Imaginary Time"
Mr. A. S. Hira (M. S. 1987)	"Cluster Modeling of Hydrogen Interaction with Lithium Surfaces"
Ms. L. Glaze (M. S. 1988) (Co-Advisor: Dr. S. Self, Department of Geology, The University of Texas at Arlington)	"A Diffusion Model Applied to Remote Sensing Studies of Volcanic Eruption Plumes"

Ms. S. Berry (M. S. 1989)	"Mass Spectra and Leptonic Decay Widths of the CC and BB Systems"
Ms. S. Altekar (M. S. 1989)	"An *Ab Initio* Study of Potassium and Mixed Pota-ssium/Magnesium Clusters"
Mr. M. Islam (D. Sc. 1989)	"An *Ab Initio* Study of Carbon and Germanium Clusters"
Mr. A. S. Hira (Ph. D. 1990)	"Many Body Perturbation Theory Applied to Hydrogen Interaction with Lithium Surfaces"
Ms. J. Tripathi (M. S. 1990)	"A Class of Quark Confinement Potentials"
Ms. K. Kanal (M. S. 1991)	"On the Existence and Stability of Potassium Clusters"
Mr. T. Kellogg (M. S. 1993)	"One Dimensional Electron Transport: Application to Electron Backscattering"
Mr. P. Lin (M. S. 1993)	"Stabilities of Adsorption Sites and Charge Transfers at the Li/Si (100) and Na/Si (100) Surfaces"
Mr. R. P. Datta (Ph. D. 1993)	"Coupled Electron Photon Transport: A Multi-Dimensional Discrete Ordinates Study"
Mr. K. M. Song (Ph. D. 1994)	"An *Ab Initio* Study of Alkali Metal Adsorption on Gallium Arsenide (110) Surface"
Ms. S. D. Altekar (Ph. D. 1995)	"Parallel Programming in Electron Transport: A Cray YMP Study"
Ms. M. Panda (M. S. 1997)	"Interactions of Alkali Atom with GaAs (110) Surface: A Density Functional Approach"
Mr. R. Schailey (Ph. D. 1999)	"An *Ab Initio* Cluster Study of Chemisorption of Atomic Cesium and Hydrogen on Reconstructed Surfaces of Gallium Rich Gallium Arsenide (100)"
Ms. X. Wu (Ph. D. 2001)	"Density Functional Theory Applied to d- and f- Electron Systems"
Mr. N. E. Frick (M. S. 2002)	"An *Ab Initio* Study of Alkali-C_{60} Complexes"

Mr. M. N. Huda (M.S. 2002)	"A Perturbation-Theoretic and Coupled-Cluster Study of Silver Nanostructures"
Mr. M. L. Mayo (M. S. 2003)	"A Many Body Perturbation Theoretic Study of Atomic Oxygen and Aluminum Chemisorption on GaAs (100) (2x1) and β (4x2) Surfaces"
Mr. M. N. Huda (Ph. D. 2004)	"A Relativistic Density Functional Study of the Role of 5f Electrons in Atomic and Molecular Adsorptions on Actinide Surfaces"
Ms. P. Pradhan (M. S. 2005)	"Nanoclusters to Nanotubes: An *Ab Initio* Study"
Ms. L. G. Sridevi (M. S. 2005)	"On a Class of ABCS"
Mr. P. Dholabhai (M. S. 2005)	"On the Behavior of 5f Electrons"
Mr. A. Srinivasan (M. S. 2005)	"On the Existence and Stability of Carbon-Based Silicon Fullerenes: A Density Functional Theoretic Study"
Mr. K. M. Alam (M. S. 2008)	"Silicon Carbide Nanotubes: Promises Beyond Carbon Nanotubes"
Mr. S. Rathi (M. S. 2008)	"Germanium Based Nanotubes: Fact or Fiction?"
Mr. P. Dholabhai (Ph. D. 2008)	"Probing the *5f* Electrons: A Relativistic DFT Study of Americium Surfaces"
Mr. S. Mukherjee (M. S. 2008)	"Can Silicon Carbide Nanotubes Be Effective Medium for Hydrogen Storage?"

Current Undergraduate Students

Ms. D. Baker
Ms. R. Pathak

Current Graduate Students

Mr. K. Adhikari
Mr. H. Chen

Ms. S. Duesman
Ms. K. Gandha
Ms. S. Hernandez
Ms. D. Morrison
Mr. P. Wannaguru

Current Post-Doctoral Fellows

Dr. Li Ma
Dr. J. Wang

Graduate Thesis/Dissertation Committees

J. Alleto (M. S.)
O. Aniltruk (Ph. D.)
S. Ayyalasomayajula (M. S.)
H. Che (M. S.)
C. Ethridge (D. Sc.)
N. Guler (M. S.)
N. Hozhabri (D. Sc.)
C. Hyer (Ph. D.)
C. Jiang (M. S.)
M. Jibaly (D. Sc.)
K. Lee (M. S.; D. Sc.)
E. Livesay (Ph. D.)
R. Mittal (M. S.; Ph. D.)
M. Morse (M.S.)
S. Mukherjee (Ph. D.)
K. O'Brien (Ph. D.)
R. Ramsey (Ph. D.)
J. Rejczek (Ph. D.)
R. Sharma (M. S.)
S. Shin (M. S.)
M. Sosbee (Ph. D.)
H. Tu (M. S.)
X. Wang (M. S.)
D. Wu (M. S.; Ph. D)
Y. Xia (M. S.)
A. Zhao(Ph. D.)

Undergraduate Students Directed

S. Anderson
M. Averitte
K. Balasubramanium
S. Bell
M. Eberwine
C. Erhorn
A. Flores
E. Gonzales
C. Harenza
A. Hendrix
K. Honza
E. James
N. Kohli
R. Minyard
A. Naqvi
L. Page
P. Pappas
A. Passi
D. Price
J. Richards
D. Wagner
K. Wright

Courses Taught

Texas Tech University	Title	Semesters
Mathematics 137,138	Mathematical Analysis for Business Majors	3
Physics 141, 142	General Physics I and II	7

University of FLorida	Title	Quarters
Physics 2050	General Physics I	4
Physics 2052	General Physics II	2
Physics 2053	Applied Physics II	1

MICHIGAN TECH. UNIVERSITY	TITLE	QUARTERS
Physics 204	Physics for Engineers I	2
Physics 209	General Physics III	2

UNIVERSITY OF TEXAS AT ARLINGTON	TITLE	SEMESTERS
Physics 1441	General College Physics I	20
Physics 1442	General College Physics II	2
Physics 1445	Introductory Astronomy I	1
Physics 3313	Modern Physics	4
Physics 4117	Individual Learning by Seminar	1
Physics 4181	Special Problems	6
Physics 4281	Special Problems	5
Physics 4326	Introduction to Quantum Mechanics	1
Physics 5190	Physics Colloquium	4
Physics 5307	Quantum Mechanics I	11
Physics 5308	Quantum Mechanics II	7
Physics 5310	Statistical Mechanics	1
Physics 5311	Mathematical Methods in Physics I	7
Physics 5312	Mathematical Methods in Physics II	5
Physics 5316	Solid State Physics II	1
Physics 5319	Mathematical Methods in Physics III	6
Physics 5393	Readings in Physics	4
Physics 5394	Research	32
Physics 5694	Research	22

Physics 5698	Thesis	20
Physics 6999	Dissertation	9
Physics 6302	Methods of Applied Physics II: Computers in Physics	1
TECHNICAL UNIVERSITY OF VIENNA	TITLE	SEMESTERS
Chemistry	Density Functional Theory	1

HONORS AND ORGANIZATIONS

Government of India National Scholar (1967).

Who's Who in Frontiers of Science and Technology (1990).

Member of the American Physical Society (1978 – cont.).

Member of American Nuclear Society (1984-87).

Member of the Materials Research Society (2005-cont.)

Member of Sigma Pi Sigma (1973-77).

Member of the Society for Industrial Applications of Mathematics (1977-1979).

Consultant, RELTEC Corporation, Dallas, Texas (1997).

Who's Who in America, 2005-2010.

College of Science Research Excellence Award, University of Texas at Arlington, 2004 -2010.

Who's Who in Science and Engineering, 2006-2008

Nominated for Distinguished Long-Term Research Award, University of Texas at Arlington, 2005, 2006, 2007.

Who's Who in the World, 2007.

Committees, Physics Department, The University of Texas at Arlington

Member of Undergraduate Studies Committee (1978-79).

Member of the Committee on Text of Modern Physics (1984-85).

Member of the Committee on Industrial Relations (1985-86).

Chair or Co-Chair of the Committee on Computer Use (1984-2002).

Member of the Committee on Faculty Search (1984-96, 2000- 2006, 2008-2009).

Member of the Committee on Undergraduate Curriculum (1985-87; 1989-90).

Member of the Graduate Studies Committee (1985-cont.).

Co-Chair of the Colloquium Committee (1986-89, 2000-01).

Co-Chair of the Graduate Recruitment Committee (1987-92).

Member of the Teaching Committee (1988-91).

Chair of the Graduate Studies Committee (1991-98).

Member of the Library Committee (1985-96, 1999-cont.).

Member of the Research Committee (1985-96, 1999-2004).

Chair of the Research Committee (2004-2006).

Chair of the Graduate Admissions Committee (1988-98, 1999-2004, 2008-2010).

Member of the Graduate Admissions Committee (2004-cont.).

Doctoral Graduate Advisor (1988-92, 2001- 2003).

Graduate Advisor (2008-2010)

Member of the Tenure and Promotion Committee (1988-96; 2000-02; 2003-2010).

Chair of the Tenure and Promotion Committee (1996-97; 2000-01).

Chair, Graduate Studies Subcommittee (1991-95).

Member of the Awards Committee (1999-2001, 2008-cont.).

Member of the Long Range Planning Committee (1999-2001).

Member of the Accommodation Committee (1997-2002).

Chairman's Committee (1997-1998; 1999-2001; 2004-cont.).

M. S. Graduate Advisor (1999-2001).

Chair, Theory Search Committee (2004).

Member of Strategic Planning Committee (2003-2008).

Member of the Astronomy Instructor Search Committee (2004-2005).

Member of the Task Force on Computer Security (2005-2006).

Chair of the Associate Chair Review Committee (2004-2005).

Chair of the Chair Review Committee (2008-2009).

Chair of the Committee for the Acquisition of a Beowulf Parallel Cluster (2005).

Member of the Computer Cluster Committee (2006-2008)

Member of the Periodic Review of Tenure Committee (2006-2008)

Member of the Periodic Review of Tenure Appeals Committee (2000-cont.)

Member of the Science Hall Renovation/Reprogramming (2005-2009)

Member of the Tenure and Promotion Guidelines Committee (2006-2009).

University Committees

Associate Member of the Graduate Faculty (1984-88).

Member of the Graduate Faculty (1988-cont.).

Member of the Undergraduate Curriculum Committee, College of Science (1989-90).

Member of the Center for High-Performance Committee User Advisory Committee, The University of Texas System (1989-94).

Member of the Ad Hoc Supercomputing Facilities Committee (1988-94; 1996-cont.).

Member of the Registration, Scheduling and Calendar Committee (1998-2000).

Member of the College of Science Tenure and Promotion Committee (2000-2002; 2004-2006, 2009-2010).

Member of College of Science Research Committee (2005-2007).

Member of the University Graduate Assistants' Task Force (2005-cont.).

Faculty Mentoring Program (2004-2005).

Chair, High Performance Computing Users Group (2008 – cont.).

UTA Representative to the Texas Advanced Computing Center User Advisory Group (2008 – cont.)

In: Superconductivity and Condensed Matter …
Editor: Collin J. Eckles

ISBN 978-1-61209-785-5

Chapter 28

BIOGRAPHICAL SKETCH – Superconductivity Research

NAME:	TITLE:	DATE OF BIRTH:
A.S.SHALABI	PROF.DR	5-8-1948

EDUCATION

Institution and Location	Degree	Year Conferred	Scientific Field
Ain Shams University, Cairo, Egypt.	B.Sc.	1977	Chemistry
Sheffield University, Sheffield, England.	Ph.D	1981	Chemistry

RESEARCH AND PROFESSIONAL EXPERIENCE

PROFESSIONAL APPOINTMENTS

HISTORY OF EMPLOYMENT

1. Assistant Professor, Department of Chemistry, Faculty of Science, Benha University, Benha, P.O.Box 13518, Egypt, 1981-1986.
2. Associate Professor, Department of Chemistry, Faculty of Science, Benha University, Benha, P.O.Box 13518, Egypt, 1986-1989.
3. Associate Professor, Department of Chemistry, Faculty of Science, UAE University, AL- Ain, P.O.Box 17551, United Arab Emirates, 1989-1995.
4. Professor, Department of Chemistry, Faculty of Science, Benha University, Benha , P.O.Box 13518, Egypt, 1993- 2002
5. Professor, Department of Chemistry, College of Science, Qatar University, Doha, P.O.Box 2713, Qatar, 2002-2005
6. Professor, Department of Chemistry, Faculty of Science, Benha University, Benha , P.O.Box 13518, Egypt, 2005-

ACADEMIC VISITS

1. Visiting Professor at School of Chemistry and Molecular Sciences, Sussex University, Falmer, Brighton, BNI 9QJ,UK, Feb.-May, 1985.
   A.S.Shalabi and J.N. Murrell
   Ab initio Clculations of Small Polyatomic Molecules.

2. Visiting Professor at Minnesota Supercomputer Institute, University of Minnesota, Twin Cities Campus, 1200 Washington Avenue South, Minneapolis, MN 55415, USA, June-August, 1991.
   A.S.Shalabi and D.G. Truhlar.
   Atom Diatom Quantum Scattering Calculations.

3. Visiting Professor at King's College London, Strand, London, WC2R, Department of Chemistry, U.K, June-August, 1992.
   A.S.Shalabi and M.A. Robb.
   Clifford Algebra Unitary Group Approximation of Electronic Structure Computations.

REFEREEING

1. J. Chem. Inf. Comput. Sciences, U.S.A
2. J. Physical Chemistry, U.S.A
3. J. American Chemical Society, U.S.A
4. Physica B. Condensed Matter Physics
5. Supreme Council of Egyptian Universities SCEU
   (The Permanent Scientific Committee, Chemistry), Egypt
6. Cairo University Awards
   (Basic Sciences, Chemistry), Egypt

MEMBERSHIPS OF SCIENTIFIC SOCIETIES AND GROUPS

1. American Chemical Society, U.S.A
2. Gaussian 82-98 users, U.S.A
3. RMPROP users, U.S.A
4. Gaussian 92W-98W users, U.S.A
5. New York Academy of Sciences, U.S.A
6. Member of Marquis Who's Who in Science and Engineering, U.S.A
7. Invited guest editor, J.Combinatorial Chemistry, U.S.A
8. Member of Innovation Explorers by Elsevier, 2010, Holand

TEACHING ACTIVITIES

UNDERGRADUATE COURSES

General Chemistry (I & II)- Quantum Chemistry- Molecular Spectroscopy-
Chemical Group Theory- Statistical Mechanics – Statistical Thermodynamics
Physical Chemistry (I : Kinetic Theory, Thermodynamics, and Phase Rule)
Physical Chemistry (II: Surface Chemistry ,Colloids and catalysis-Chemical kinetics) – Sloid State Chemistry- Photo Chemistry (laser)

POSTGRADUATE COURSES

Quantum Chemistry- Molecular Spectroscopy-Computational Chemistry.
Chemical Group Theory

RESEARCH ACTIVITIES

Chemical applications of Quantum, Group and Graph Theories – Molecular Spectroscopy - Solid state theory – Adsorption – Laser - Photography.

HONORS RELEVANT TO RESEARCH

SUPERVISION OF M.SC AND PH.D STUDENTS

14 M.Sc. Theses
11 Ph.D Theses

REFEREEING

1. J. Chem. Inf. Comput. Sciences, U.S.A
2. J. Physical Chemistry, U.S.A
3. J. American Chemical Society, U.S.A
4. Supreme Council of Egyptian Universities SCEU
   (The Permanent Scientific Committee, Chemistry), Egypt
5. Cairo University Awards
   (Basic Sciences, Chemistry), Egypt

MEMBERSHIPS OF SCIENTIFIC SOCIETIES AND GROUPS

1. American Chemical Society, U.S.A
2. Gaussian 82-98 users, U.S.A
3. RMPROP users, U.S.A

4. Gaussian 92W-98W users, U.S.A
5. New York Academy of Sciences, U.S.A
6. Member of Marquis Who's Who in Science and Engineering,U.S.A
7. Invited guest editor, J.Combinatorial Chemistry,U.S.A
8. Member of Innovation Explorers by Elsevier, 2010, Holand

Publications during Last Three Years

W.S.Abdel Halim, S.Abdel Aal, **A.S.SHALABI**, CO adsorption on Pd atoms deposited on MgO, CaO, SrO and BaO: Density functional calculations, Thin Solid Films 516 (2008) 4360

A.S.Shalabi *, N.Abdullah, W.S.Abdel Halim,Kh.M.Eid, M.A.Kamel, H.O.Taha, Optical properties of charged* F^+A_1*:*Zn^{2+}*,*Cd^{2+}*,*Hg^{2+} *centers in the bulk and at the surface of MgO : Theoretical study,* Physica B 403 (2008) 3168

W.S.Abdel Halim, **A.S.Shalabi**, K.A.Soliman, Transition metal atoms on oxide supports: Density functional calculations, I.J.Quantum Chemistry 109 (2009) 1094

S.Abdel Aal, W.S.Abdel Halim, **A.S.Shalabi**, Cl_2 adsorption on supported alkali metals and on the MgO and CaO (001) supports: A DFT study, Solid State Communications 148 (2008) 464

W.S.Abdel Halim, **A.S.Shalabi,** K.A.Soliman, Transition Metal Atoms on Oxide Supports: Density Functional Calculatons, I.J.Quantum Chemistry 109(2009)1049

W.S.Abdel Halim, M.M.Assem, **A.S.Shalabi,** K.A.Soliman, CO adsorption on Ni, Pd, Cu, and Ag deposited on MgO, CaO, SrO, and BaO, Density Functional Calculations, Appl. Surf. Sci. 255(2009)7574.

A.S.Shalabi, W.S.Abdel Halim, M.S.Ghonaim, Spin quenching of Transition Metals deposited on MgO insulator and CdO semiconductor: Density Functional Calculations, I.J.Quantum Chemistry (2009) accepted

Other Contributions of Interest to Research: Projects, Grants etc.

1. British Council Ph. D. Fees Support Scheme, 1978, U.K
2. British Council Postdoctor Research Fellowship, 1985, U.K
3. British Council Postdoctor Research Fellowship, 1992, U.K

In: Superconductivity and Condensed Matter …
Editor: Collin J. Eckles
ISBN 978-1-61209-785-5

Chapter 29

BIOGRAPHICAL SKETCH – Superconductivity Research

NAME:	TITLE:	DATE OF BIRTH:
NATHALIE OLIVI-TRAN	DR.	DECEMBER 27TH 1967

EDUCATION

Institution and Location	Degree	Year Conferred	Scientific Field
French Lycée B. Pascal, Abidjan	Baccalaureate	1985	Mathematics and Physics
Lycée Saint-Louis, Paris	Preparation to competetive exams	1985-87	Mathematics
Medicine Faculty (French Army,Lyon)	PCEM1-2	1987-89	Medicine
UniversityMontpellier II	Master, PhD	1995	Physics
University of Amiens	Assistant Professor short term	1995-96	Physics
University of Paris-Sud, Orsay	Postdoctorate	1996-97	Physics
University of Nice	Habilitation à diriger des recherches	2003	Physics

RESEARCH AND PROFESSIONAL EXPERIENCE:

PROFESSIONAL APPOINTMENTS

2009-present: Junior scientist (Chargee de recherches 1ere classe (CNRS)) in the Groupe d'Etude des Semiconducteurs UMR-CNRS-5650, France

2006-2008: Scientific stay in the Laboratoire de Physique de la Matière Condensée et Nanoparticules (Villeurbanne, University Lyon I)

2004-2008: Junior scientist (Chargee de recherches 1ere classe (CNRS)) in the Laboratoire de Science des Procedes et Traitements de Surface UMR-CNRS-6638, France

2001-2004: Junior Scientist (Chargee de recherches 1ere classe(CNRS)) in the Institut Non Lineaire de Nice (UMR 6618), France

1997-2001: Junior scientist (Chargee de recherches 2eme classe (CNRS)) in the Institut Non Lineaire de Nice (UMR 6618), France

Publications during Three Years

- O.Pozo, B.Soulestin and **N.Olivi-Tran**, *Stick Slip in grain grain friction in a humid atmopshere,* Multidiscipline Modeling in Materials and Structures(MMMS) Vol.4 no.4 (2008) 393-405(ArXiv:cond-mat/0510532).
- O.Pozo and **N.Olivi-Tran**, *Effects of elasticity on the shape of measured shear signals in a two-dimensional assembly of disks* Multidiscipline Modeling in Materials and Structures(MMMS) vol.4 no.2 (2008) pp.179-188
- F.Lallet, **N.Olivi-Tran** and A.Dauger, *Experiments and simulations of thin solid films islanding* book chapter in ' Ceramics Materials: Research Trends' Editor Paul B.Lin Nova Science Publishers Inc. New York 2007
- **N.Olivi-Tran** and P.M.Gauthier, *The FLRW cosmological model revisited: Relation on the local time with the local curvature and consequences on the Heisenberg uncertainty principle* Adv. Studies Theor. Phys. vol.2 no 6 (2008) 267-270 (published online, freely available at www.m-hikari.com)
- F.Lallet, **N.Olivi-Tran** and L.J.Lewis *Interface energies of (100)_{YSZ} and (111)_{YSZ} epitaxial islands on (0001)_{alpha-Al_2O_3} substrates from first principles*Phys. Rev. B 79 (2009) 035413 (12 pages) (ArXiv:0801.3594v1)
- K.Hsieh, F.Lallet and **N.Olivi-Tran** *DLCA and Langevin Dynamics Approaches of Sol Gel Transition: a comparison via the fractal dimension* FRACTALS Vol.16 issue 4 (2008)361-365
- **N.Olivi-Tran**, S. Kanani, I.Robbins *The role of the extra cellular matrix on memory* Advanced Studies in Biology, Vol.1, no 1 (2009) 9-14 (published online, freely available at www.m-hikari.com) [hal-00134775 - version 2] (2007-03-22)
- **N.Olivi-Tran** *Surface fraction of random close packing in two dimensions calculated exactly*Adv. Studies in Theo. Phys. Vol. 2, 2008, no. 20, 975 - 978(published online,freely available at www.m-hikari.com ; arXiv:cond-mat/0808.4070)
- M.Leleyter and **N.Olivi-Tran** *Tight-Binding Study of Si2Cn (n=3 to 42) Fullerene-like or Nanodimaonds Microclusters: are Si atoms isolated or adjacent?* Eur. Phys. J. D Vol.50 (2008) 153-163
- **N.Olivi-Tran**,S.Calas,P.Etienne *Connectivity of hybrid polymer by numerical simulation: example of $(Si(CH_3)_2)_n(OSiO)_p(OH)_q$* J. of Polymers and the Env. vol.17 (2009) pp.187-190 (arXiv:0810.1185)
- **N.Olivi-Tran**,A.Ferchichi, S.Calas, P.Etienne, *Total electronic energy by tight binding approximation and experimental toughness of three different hybrid polymers* J. of Non-Cryst. Solids Vol.356 (2010) 287-289(arXiv:0811.1487)
- **N.Olivi-Tran** *Dimensional analysis of Einstein's fields equations* Adv. Studies Theor. Phys., Vol. 3, 2009, no. 1, 9 - 12 (published online,freely available at www.m-hikari.com)

- **N.Olivi-Tran** and A.Faivre *Analysis of a Lennard-Jones fcc structure melting to the corresponding frozen liquid: differences between the bulk and the surface* Physica B 404 (2009) 2289-2293 (arXiv:0902.4101)
- F.Lallet, **N.Olivi-Tran** *Sol-gel obtained YSZ nanocrystals on a substrate: a Monte Carlo approach of the crystallographic orientation* J.of Luminescence vol.129 (2009) pp.1767-1770 (arXiv:0902.4104)
- **N.Olivi-Tran** *Numerical study of the antiferromagnetic Ising model on a hypersphere* Adv. Studies Theor. Phys., Vol. 3, 2009, no. 12, 481 - 488 (published online,freely available at www.m-hikari.com;ArXiv:physics/0504041).
- J.R.Darias and **N.Olivi-Tran** *Two dimensional curved disks on a sphere: the evolution of kinetic energy*Adv. Theor. Appl. Mech., Vol. 2, 2009, no. 4, 159 - 165 (published online,freely available at www.m-hikari.com;arXiv:cond-mat/0702013
- **N.Olivi-Tran** *What if our three dimensional curved universe was embedded in four dimensional space? Consequences on the EPR paradox*Adv. Studies Theor. Phys., Vol. 3, 2009, no. 12, 489 - 492,(published online,freely available at www.m-hikari.com)
- **N.Olivi-Tran** *Pure, Si and sp3-C-doped graphene nanoflakes: A numerical study of density of states* in press Physica B (arXiv:1002.1888)
- **N.Olivi-Tran** *Is it the Higgs scalar field?* Advanced Studies in Theoretical Physics Vol. 4, 2010, no. 13, 633 - 636 (forthcoming article; www.m-hikari.com; open access journal)

OTHER CONTRIBUTIONS OF INTEREST TO RESEARCH: PROJECTS, GRANTS ETC.

1. Researchers exchanges: Convention d'échanges CNRS-FONACIT1998 (France-Vénézuela)
2. Researchers exchanges CNRS-FONACIT 1999 (France-Vénézuela)
3. Research Grant: Action Concertée Incitative Jeunes Chercheurs 'Structures Localisées' 1999-2003
4. Postgraduate students exchange: Programme de Coopération Postgraduée (France-Vénézuela) 2005-2008
5. Researchers exchanges: Convention d'échanges CNRS-FONACIT 2004-2006 (France-Vénézuela)

In: Superconductivity and Condensed Matter ...
Editor: Collin J. Eckles
ISBN 978-1-61209-785-5

Chapter 30

BIOGRAPHICAL SKETCH – Superconductivity Research

NAME:	TITLE:	DATE OF BIRTH:
ADITYA MAHABHAI VORA	DR.	FEBRUARY 27TH 1974

M. SC., M. PHIL., PH. D. (PHYSICS), B. ED., C.I.C., M. A. (EDU.-CONT.)

PERSONAL DETAILS

Permanent Address (✉)	Parmeshwari, Plot No. – 165, Vijaynagar Area, Hospital Road, Bhuj-Kachchh 370001, Gujarat, INDIA
Present Address (✉)	I/C Head, Humanities and Social Science Department, STBS College of Diploma Engineering, Opp. Spinning Mill, Varachha Road, Surat 395006, South Gujarat, India
Contact Numbers (☎)	(R) 091–2832–256424 (Mob.) 091–9426956016
E-mail Address (✉)	voraam@yahoo.com, voraam@rediffmail.com, voraam@gmail.com
Birth Date & Birth Place	27th February, 1974 Bhuj-Kachchh
Passport Number	E 6786220 (Exp. Date : 28/09/2013)
Marital Status	Married
Gender	Male
Nationality	Indian
Languages Known	English, Hindi, Gujarati
Employment Exchange Registration Number	200400871, University Employment Bureau, Saurashtra University, Rajkot

ACADEMIC RECORD

DEGREE/ EXAM	Year	BOARD / UNIVERSITY	CLASS %	SUBJECT
S. S. C.	1989	GSEB	FIRST 64%	SCIENCE
H. SC.	1991	GHSEB	PASS 54%	SCIENCE
B. Sc.	1995	Gujarat	Second 66%	Physics
M. Sc.	1997	Sardar Patel	Second 58%	Physics (Solid State Electronics)
M. Phil.	1998	Sardar Patel	First 65%	Physics (Experimental Condensed Matter Physics)
Ph. D.	2004	Sardar Patel	Physics (Theoretical Condensed Matter & Computational Physics)	
B. Ed.	2006	Kachchh	Dist. 84%	Science, Maths Methods
C. I. C.	2007	IGNOU	First 61%	Computer Fundamentals
M. A.	-	IGNOU	-	Education (Continue..)

LIST OF PUBLICATIONS

[1] Effect of Indium intercalation on various properties of In_xMoSe_2 ($0 \leq x \leq 1$) single crystals ***A. M. Vora***, M. P. Deshpande and P. K. Garg, Prajna-Journal of Sardar Patel University (India) **9** (1999) 47-50.

[2] The optical absorption in single crystals of In_xMoSe_2 ($0 \leq x \leq 1$), ***A. M. Vora*** and P. K. Garg, Indian Journal of Physics (India) **74A** (6) (2000) 563-566.

[3] Fermi energy and Fermi surface of Cs-K, Cs-Rb and Rb-K binary systems, M. H. Patel, ***A. M. Vora***, P. N. Gajjar and A. R. Jani, Physica B: Condensed Matter (Netherlands) **304** (2001) 152-158.

[4] Residual resistivity of some dilute alloys, ***A. M. Vora***, M. H. Patel, P. N. Gajjar and A. R. Jani Chinese Journal of Physics (Republic of China) **39** (3) (2001) 274-281.

[5] Some static and vibrational properties of equiatomic sodium based binary alloys P. N. Gajjar, ***A. M. Vora***, M. H. Patel and A. R. Jani, Different Disordered Systems, Eds. Furukawa *et al.*, INDIAS Publications, Allahabad (India) (2000) 57-61.

[6] Monovacancy resistivity of some simple metals and their alloys, ***A. M. Vora***, M. H. Patel, B. Y. Thakore, P. N. Gajjar and A. R. Jani, Solid State Physics (India) **43** (2000) 480-481.

[7] Phonon dispersion for Na-Li, Na-K, Na-Rb and Na-Cs equiatomic solid solutions

J. M. Panchal, M. H. Patel, ***A. M. Vora***, P. N. Gajjar and A. R. Jani, Proceedings of COMPOSE 2001, S. P. University, Vallabh Vidyanagar (India) (2001) 27.

[8] Superconducting state parameters of some metallic glasses, ***A. M. Vora***, M. H. Patel, S. R. Mishra, P. N. Gajjar and A. R. Jani, Solid State Physics (India) **44** (2001) 345-346.

[9] Superconducting state parameters of Indium based binary alloys, ***A. M. Vora***, M. H. Patel, P. N. Gajjar and A. R. Jani, Pramana-J. of Physics (India) **58** (5-6) (2002) 849-853.

[10]Asphericity in the Fermi surface and Fermi energy of Na-K, Na-Rb and Na-Cs binary alloys, M. H. Patel, ***A. M. Vora***, P. N. Gajjar and A. R. Jani, Communications in Theoretical Physics (China) **38** (2002) 365-369.

[11]Electrical resistivity of Na-based alkali binary alloys, ***A. M. Vora***, M. H. Patel, B. Y. Thakore, P. N. Gajjar and A. R. Jani, Solid State Physics (India) **45** (2002) 427-428.

[12]Superconducting state parameters of metals, ***A. M. Vora***, P. N. Gajjar and A. R. Jani, Proceedings of 4th National Symposium of Research Scholars (NSRS) on Metals and Materials, Indian Institute of Metals, IIT, Chennai (India) {Paper No. 38- in CD format) (2002).

[13]Electrical resistivity of K-Rb and K-Cs liquid binaries, ***A. M. Vora***, P. N. Gajjar and A. R. Jani, Proceedings of Seminar on Theoretical Physics *for* Condensed Matter, Atoms-Molecules & Particles, S. P. University, Vallabh Vidyanagar (India) (2003) 10.

[14]Vibrational dynamics of $Mg_{70}Zn_{30}$ metallic glass, ***A. M. Vora***, M. H. Patel, P. N. Gajjar and A. R. Jani, Solid State Physics (India) **46** (2003) 315-316.

[15]The study of superconducting state parameters of In–based binary alloys by pseudopotential, P. N. Gajjar, ***A. M. Vora***, M. H. Patel and A. R. Jani, International Journal of Modern Physics (Singapore) **B17** (31-32) (2003) 6001-6017.

[16]Pseudopotential in the study of superconducting state parameters of metallic glasses, P. N. Gajjar, ***A. M. Vora*** and A. R. Jani, Modern Physics Letters (Singapore) **B18** (12-13) (2004) 573-582.

[17]Superconducting state parameters of metals, P. N. Gajjar, ***A. M. Vora*** and A. R. Jani , Indian Journal of Physics (India) **78** (8) (2004) 775-780.

[18]Study of vibrational dynamics of binary metallic glasses using pseudopotential theory, ***A. M. Vora,*** Solid State Physics (India) **49** (2004) 63-64.

[19]Vibrational dynamics of $Ca_{70}Mg_{30}$ metallic glass, P. N. Gajjar, ***A. M. Vora*** and A. R. Jani, Proceedings of the 9th Asia Pacific Physics Conference Hanoi, Vietnam (October 25-31, 2004), The Gioi Publication, Vietnam (2006) 429-432.

[20]Anisotropy of the optical absorption in layered single crystals, Mihir M. Vora and ***Aditya M. Vora***, Crystal Research Technology (Germany) **41** (8) (2006) 803-806.

[21]Phonon dispersion in $Ca_{70}Mg_{30}$ metallic glass, ***Aditya M. Vora***, Chinese Physics Letters (UK) **23** (7) (2006) 1872-1875.

[22]Pseudopotential in the study of phonon dynamics of Pd-based metallic glasses, ***Aditya M. Vora,*** Journal of Non-Crystalline Solids (Netherlands) **352** (2006) 3217-3223.

[23]Study of driving forces for atomic migration in metals and dilute alloys, ***Aditya M. Vora,*** Journal of Optoelectronics and Advanced Materials (Romania) **8** (5) (2006) 1848-1850.

[24]Electrical resistivity of K-based liquid binaries, ***Aditya M. Vora,*** Russian Physics Journal (Russia) **49** (8) (2006) 834-839.

[25]Study of superconducting state parameters of alkali-alkali binary alloys by a pseudopotential, ***Aditya M. Vora***, Physica C: Superconductivity and its Applications (Netherlands) **450** (2006) 135-145.

[26]Anisotropy of the optical absorption in layered single crystals of $MoRe_{0.001}Se_{1.999}$, Mihir M. Vora and ***Aditya M. Vora,*** Crystal Research Technology (Germany) **42** (1) (2007) 50-53.

[27]Intercalation and anisotropy of optical absorption of $MoRe_{0.005}Se_{1.995}$ single crystal, Mihir M. Vora and ***Aditya M. Vora***, Crystal Research Technology (Germany) **42** (2) (2007) 186-189.

[28]Effect of indium intercalation on various properties of $MoSe_2$ single crystals, ***Aditya M. Vora***, Crystal Research Technology (Germany) **42** (3) (2007) 286-289.

[29]Residual resistivity of simple metals and their dilute alloys, ***Aditya M. Vora,*** Optoelectronics and Advanced Materials-Rapid Communications (Romania) **1** (2) (2007) 61-64.

[30]Computation of phonon dynamics of $Mg_{70}Zn_{30}$ metallic glass, ***Aditya M. Vora,*** Journal of Materials Science (Netherlands) **42** (2007) 935-940.

[31]Forces driving atomic migration in metals and dilute alloys, ***Aditya M. Vora,*** Optoelectronics and Advanced Materials-Rapid Communications (Romania) **1** (4) (2007) 182-185.

[32]Electrical resistivity of Li-based liquid binary mixtures, ***Aditya M. Vora,*** Optoelectronics and Advanced Materials-Rapid Communications (Romania) **1** (5) (2007) 227-230.

[33]Screening dependence superconducting state parameters of Cu-Zr binary metallic glasses, ***Aditya M. Vora,*** Superconductor Science and Technology (UK) **20** (2007) 542-548.

[34]Study of monovacancy resistivity of some metallic elements and their dilute alloys by pseudopotential, ***Aditya M. Vora,*** Computational Materials Science (Netherlands) **40** (1) (2007) 73-80.

[35]Superconducting state parameters of amorphous metals, ***Aditya M. Vora,*** Physica C: Superconductivity and its Applications (Netherlands) **458** (1-2) (2007) 21-28.

[36]Superconducting state parameters of Cu_xZr_{100-x} binary metallic glasses, ***Aditya M. Vora,*** Physica C: Superconductivity and its Applications (Netherlands) **458** (1-2) (2007) 43-50.

[37]Superconducting state parameters of Cu-Zr metallic glasses, ***Aditya M. Vora,*** Journal of Optoelectronics and Advanced Materials (Romania) **9** (8) (2007) 2498-2503.

[38]Phonon dispersion in amorphous Ni-alloys, ***Aditya M. Vora,*** Acta Physica Polonica A (Polend) **111** (6) (2007) 859-871.

[39]Screening dependence of superconducting state parameters of metals, ***Aditya M. Vora,*** Physica Scripta (UK) **76** (2007) 204-209.

[40]Study of superconducting state parameters of binary metallic glasses by pseudopotential method, ***Aditya M. Vora,*** Journal of Superconductivity and Novel Magnetism **20** (5) (Netherlands) (2007) 355-368.

[41]Study of screening dependence superconducting state parameters of Cu_CZr_{100-C} binary metallic glasses by a pseudopotential, ***Aditya M. Vora,*** Journal of Superconductivity and Novel Magnetism **20** (5) (Netherlands) (2007) 373-380.

[42]Superconducting state parameters of 5d-traansition metals based binary alloys , ***Aditya M. Vora,*** Journal of Superconductivity and Novel Magnetism **20** (5) (Netherlands) (2007) 387-392.

[43]Electrical properties measurements of In_xMoSe_2 ($0 \leq x \leq 1$) single crystal, Mihir M. Vora and ***Aditya M. Vora,*** Chalcogenide Letters (Romania) **4** (7) (2007) 77-83.

[44]Intrinsic stacking fault in single crystal of In_xMoSe_2 ($0 \leq x \leq 1$), Mihir M. Vora and ***Aditya M. Vora,*** Chalcogenide Letters (Romania) **4** (7) (2007) 85-88.

[45]Study of superconducting state parameters of Cu_CZr_{100-C} binary amorphous alloys by pseudopotential approach, ***Aditya M. Vora,*** Chinese Physics Letters (UK) **24** (9) (2007) 2624-2627.
[46]Static and vibrational properties of equiatomic Na-based binary alloys, ***Aditya M. Vora,*** Journal of Physics and Chemistry of Solids (UK) **68** (2007) 1725-1729.
[47]Superconducting state parameters of binary metallic glasses, ***Aditya M. Vora,*** Computational Materials Science (Netherlands) **40** (2007) 492-503.
[48]Superconducting state parameters of Cu_CZr_{100-C} metallic glasses, ***Aditya M. Vora,*** Frontiers of Physics in China (China) **2** (4) (2007) 430-435.
[49]Study of driving forces for atomic migration in dilute alloys, ***Aditya M. Vora,*** Moroccan Journal of Condensed Matter (Morocco) **9** (1) (2007) 35-66.
[50]Residual resistivity of some metallic elements, ***Aditya M. Vora,*** Moroccan Journal of Condensed Matter (Morocco) **9** (1) (2007) 108-112.
[51]Study of driving forces for atomic migration in dilute alloys, ***Aditya M. Vora,*** Moroccan Journal of Condensed Matter (Morocco) **9** (1) (2007) 121-141.
[52]Vibrational dynamics of non-crystalline glassy alloys, ***Aditya M. Vora,*** Frontiers of Materials Science in China (China) **1** (4) (2007) 366-378.
[53]Effect of rhenium doping on various properties of $MoSe_2$ single crystal, M. M. Vora and ***A. M. Vora,*** Chalcogenide Letters (Romania) **4** (8) (2007) 97-100.
[54]Electrical transport properties of liquid Cd-binary alloys, ***Aditya M. Vora,*** Journal of Ovonic Research (Romania) **3** (3) (2007) 57-66.
[55]Electrical properties of liquid semiconductor alloys, ***Aditya M. Vora,*** Journal of Ovonic Research (Romania) **3** (5) (2007) 83-91.
[56]Structure studies of alkaline-earth metals: Evaluation through the charged hard sphere approximation, ***Aditya M. Vora,*** Optoelectronics and Advanced Materials-Rapid Communications (Romania) **1** (11) (2007) 639-643.
[57]Electrical resistivity of liquid $Rb_{1-X}Cs_X$ binary alloys, ***Aditya M. Vora,*** Journal of Ovonic Research (Romania) **3** (6) (2007) 129-133.
[58]Study of superconducting state parameters of binary Be_cAl_{100-c} metallic glasses, ***Aditya M. Vora,*** Journal of Technical Physics (Poland) **48** (3-4) (2007) 3-10.
[59]Electrical resistivity of liquid alkali Na-based binary alloys, ***Aditya M. Vora,*** Turkish Journal of Physics (Turkey) **31** (2007) 341-346.
[60]Computation of phonon dispersion in non-crystalline Mg-based alloys, ***Aditya M. Vora,*** FIZIKA A (Croatia) **16** (4) (2007) 187-206.
[61]Effect of rhenium doping on various properties of $MoRe_{0.001}Se_{1.999}$ single crystal, ***Aditya M. Vora,*** Chalcogenide Letters (Romania) **5** (2) (2008) 17-20.
[62]Phonon dispersion in equiatomic Li-based binary alloys, ***Aditya M. Vora,*** Chinese Physics Letters (UK) **25** (2) (2008) 654-657.
[63]A pseudopotential approach to the superconducting state properties of Cu_CZr_{100-C} binary metallic glasses, ***Aditya M. Vora,*** Journal of Contemporary Physics (Armenian Academy of Science) (Armenia) **43** (1) (2008) 27-36.
[64]Residual resistivity of dilute alloys, ***Aditya M. Vora***, International Journal of Modern Physics B**22** (5) (Singapore) (2008) 517-527.
[65]Electrical transport properties of liquid alloys, ***Aditya M. Vora,*** Physics and Chemistry of Liquids (UK) **46** (2) (2008) 223-232.

[66]Superconducting state parameters of binary metallic glasses, ***Aditya M. Vora,*** Central European Journal of Physics (Poland) **6** (2) (2008) 238-252.

[67]Superconducting state parameters of $La_{100-C}Ga_C$ binary metallic glasses, ***Aditya M. Vora,*** Central European Journal of Physics (Poland) **6** (2) (2008) 253-262.

[68]Study of superconducting state parameters of amorphous metals by a pseudopotential theory, ***Aditya M. Vora,*** Central European Journal of Physics (Poland) **6** (2) (2008) 263-276.

[69]Thermodynamics of liquid d- and f-shell metals: A variational approach, ***Aditya M. Vora,*** Physics and Chemistry of Liquids (UK) **46** (3) (2008) 278-286.

[70]Study of vibrational dynamics of Ca-based metallic glasses by pseudopotential theory, ***Aditya M. Vora,*** Romanian Journal of Physics (Romania) **53** (3-4) (2008) 517-533.

[71]Superconducting state parameters of 3d-transition metals binary alloys, ***Aditya M. Vora,*** International Journal of Theoretical Physics, Group Theory and Nonlinear Optics (New York) **12** (4) (2008) 283-294.

[72]Superconducting State Parameters of Pb-Tl-Bi Alloys, ***Aditya M. Vora,*** Journal of Ovonic Research (Romania) **4** (1) (2008) 13-19.

[73]Stacking faults in Re doped $MoSe_2$ single crystals, Mihir M. Vora and ***Aditya M. Vora,*** Chalcogenide Letters (Romania) **5** (3) (2008) 35-37.

[74]Effect of rhenium doping on various physical properties of single crystals of $MoSe_2$, Mihir M. Vora and ***Aditya M. Vora,*** Chalcogenide Letters (Romania) **5** (4) (2008) 65-71.

[75]Modified transition temperature equation for superconductors, ***Aditya M. Vora,*** Chinese Physics Letters (UK) **25** (6) (2008) 2162-2164.

[76]Screening dependence superconducting state parameters of $La_{100-C}Ga_C$ binary metallic glasses, ***Aditya M. Vora,*** Journal of Physics and Chemistry of Solids (UK) **69** (7) (2008) 1841-1847.

[77]Superconductivity of amorphous $Mg_{0.70}Zn_{0.30-x}Ga_x$ alloys, ***Aditya M. Vora,*** Physica C: Superconductivity and its Applications (Netherlands) **468** (11-12) (2008) 937-943.

[78]Superconducting state parameters of Be-Zr glassy alloys, ***Aditya M. Vora,*** International Journal of Theoretical Physics, Group Theory and Nonlinear Optics (New York) **13** (1) (2008) 1-12.

[79]Study of vibrational dynamics of $Pd_{77.5}Si_{16.5}Cu_6$ bulk metallic glass, ***Aditya M. Vora,*** International Journal of Theoretical Physics, Group Theory and Nonlinear Optics (New York) **13** (1) (2008) 31-50.

[80]Thermodynamics of liquid metals: A variational approach, ***Aditya M. Vora,*** International Journal of Theoretical Physics, Group Theory and Nonlinear Optics (New York) **13** (1) (2008) 67-85.

[81]Knight-shifts in liquid non-transition and transition metals: a pseudopotential approach, ***Aditya M. Vora,*** Chinese Journal of Physics (Republic of China) **46** (4) (2008) 426-437.

[82]Study of superconducting state parameters of ternary metallic glasses through pseudopotential approach, ***Aditya M. Vora,*** Science and Technology of Advanced Materials (UK) **9** (2008) 025017.

[83]Theoretical investigation of superconducting state parameters of binary metallic glasses $Cu_{1-c}Sn_c$, ***Aditya M. Vora,*** Turkish Journal of Physics (Turkey) **32** (2008) 199-210.

[84]Superconducting state parameters of Be-Zr glassy alloys, ***Aditya M. Vora,*** Turkish Journal of Physics (Turkey) **32** (2008) 219-226.

[85]Electrical transport properties of some liquid metals, ***Aditya M. Vora,*** Physics and Chemistry of Liquids (UK) **46** (4) (2008) 442-453.

[86]Superconducting state parameters of Pb-Tl-Bi alloys, ***Aditya M. Vora,*** Technical Physics Letters (Russia) **34** (9) (2008) 740-742.

[87]Static and vibrational properties of alkali metals, ***Aditya M. Vora,*** Moroccan Journal of Condensed Matter (Morocco) **10** (1) (2008) 7-14.

[88]Superconducting state parameters of ternary metallic glasses, ***Aditya M. Vora,*** Moroccan Journal of Condensed Matter (Morocco) **10** (1) (2008) 15-21.

[89]Phonon dispersion in alkali metals and their equiatomic sodium based binary alloys, ***Aditya M. Vora,*** Frontiers of Materials Science in China (China) **2**(3) (2008) 311-315.

[90]Superconducting state parameters of Ag_xZn_{1-x} and Ag_xAl_{1-x} binary alloys superconductors, ***Aditya M. Vora,*** Journal of Shanghai University (English Edition) (China) **12** (4) (2008) 311-317.

[91]Superconducting state parameters of binary alloys, ***Aditya M. Vora,*** Romanian Journal of Physics (Romania) **53** (7-8) (2008) 885-895.

[92]Screening-dependent superconducting state parameters of transition metals-based binary alloys, ***Aditya M. Vora,*** Journal of Contemporary Physics (Armenian Academy of Science) **43** (5) (Armenia) (2008) 231-237.

[93]Pressure dependence of the superconducting state parameters of binary $Ca_{70}Mg_{30}$ metallic glass superconductor, ***Aditya M. Vora,*** Journal of Contemporary Physics (Armenian Academy of Science) **43** (6) (Armenia) (2008) 293-300.

[94]Pseudopotential in the study of superconducting state parameters of metallic glasses, ***Aditya M. Vora,*** Journal of Non-Crystalline Solids (Netherlands) **354** (45-46) (2008) 5022-5032.

[95]Superconducting state parameters of ternary amorphous superconductors, ***Aditya M. Vora,*** Physica C: Superconductivity and its Applications (Netherlands) **468** (22) (2008) 2292-2298.

[96]Phonon dispersion in binary metallic glasses, ***Aditya M. Vora,*** Glass Physics and Chemistry (Russia) **34** (6) (2008) 671-682.

[97]Computational study of vibrational dynamics of binary $Mg_{0.70}Zn_{0.30}$ metallic glass by a pseudopotential theory, ***Aditya M. Vora,*** Turkish Journal of Physics (Turkey) **32** (2008) 323-330.

[98]Lattice dynamics of equiatomic binary alloys, ***Aditya M. Vora,*** Fizika A (Croatia) **17** (3) (2008) 87-102.

[99]Electrical transport properties of some liquid metals, ***Aditya M. Vora,*** High Temperature (Russia) **46** (6) (2008) 800-810.

[100] Metallic glasses : a brief review, ***Aditya M. Vora,*** Materials Science Research Journal (New York) **2** (1-2) (2008) 19-24.

[101] Superconducting state parameters of Al-Li binary alloys, ***Aditya M. Vora,*** Turkish Journal of Physics (Turkey) **33** (2009) 57-62.

[102] Study of superconducting state parameters of alloy superconductors, ***Aditya M. Vora,*** Electronic Journal of Theoretical Physics **6** (20) (2009) 371-380.

[103] Study of magnetic susceptibilities in liquid metals, ***Aditya M. Vora,*** Physics and Chemistry of Liquids (UK) **47** (1) (2009) 61-73.

[104] Study of superconducting state parameters of transition metals binary alloys by a pseudopotential approach, ***Aditya M. Vora,*** Modern Physics Letters B (Singapore) **23** (2) (2009) 217-227.

[105] Screening dependence study of superconductivity in 3d-transition metals binary alloys superconductors, ***Aditya M. Vora,*** Communications in Theoretical Physics (China) **51** (3) (2009) 533-539.

[106] Transport properties of Si and Ge liquid semiconductor metals, ***Aditya M. Vora,*** Communications in Theoretical Physics (China) **51** (3) (2009) 550-554.

[107] Theoretical study of superconducting state parameters of binary metallic glasses by a pseudopotential method, ***Aditya M. Vora,*** Physica C: Superconductivity and its Applications (Netherlands) **469** (2009) 241-246.

[108] Pressure effects on $Mg_{70}Zn_{30}$ superconductors, ***Aditya M. Vora,*** Modern Physics Letters B (Singapore) **23** (11) (2009) 1443-1455.

[109] Thermodynamic properties of liquid metals using Percus-Yevick (PY) hard sphere reference system, ***Aditya M. Vora,*** Digest Journal of Nenomaterials and Biostructures (Romania) **4** (1) (2009) 113-118.

[110] Vibrational dynamics of $Co_{67}Zr_{33}$ semiconductor glassy alloy, ***Aditya M. Vora,*** Journal of Structural Chemistry (Russia) **50** (2) (2009) 219-227.

[111] Vibrational dynamics of binary metallic glasses, ***Aditya M. Vora,*** Journal of Non-Oxide Glasses (Romania) **1** (2) (2009) 157-173.

[112] Study of superconducting state parameters of $Nb_xTa_yMo_z$ superconductors by pseudopotential theory, Aditya M. Vora, Armenian Journal of Physics (Armenian Academy of Science) (Armenia) **2** (3) (2009) 213-223.

[113] The study of superconducting state parameters of $In_{1-X}Tl_X$ binary alloys by pseudopotential theory, ***Aditya M. Vora,*** High Temperature (Russia) **47** (5) (2009) 635-642.

[114] Vibrational Dynamics of Zr-based bulk metallic glasses, ***Aditya M. Vora,*** Frontiers of Materials Science in China (China) **3** (3) (2009) 285-300.

[115] Thermodynamic properties of liquid metals using Percus-Yevick (PY) hard sphere reference system, ***Aditya M. Vora,*** Journal of Engineering Physics and Thermophysics (Belarus) **82** (4) (2009) 779-788.

[116] Stacking faults in the single crystals, Mihir M. Vora and ***Aditya M. Vora,*** Semiconductor Physics, Quantum Electronics and Optoelectronics **12** (4) (Ukrain) (2009) 421-423.

[117] Study of electron dispersion curves in liquid alkalis, ***Aditya M. Vora,*** Physics and Chemistry of Liquids (UK) **47** (6) (2009) 663-672.

[118] Study of vibrational dynamics of $Fe_{80}B_{14}Si_6$ bulk metallic glass, ***Aditya M. Vora,*** International Journal of Theoretical Physics, Group Theory and Nonlinear Optics (New York) **13** (2) (2009) – in press.

[119] Superconducting state parameters of bulk metallic glasses, ***Aditya M. Vora,*** International Journal of Theoretical Physics, Group Theory and Nonlinear Optics (New York) **13** (2) (2009) – in press.

[120] Study of vibrational dynamics of binary $Mg_{70}Zn_{30}$ metallic glass, ***Aditya M. Vora,*** International Journal of Theoretical Physics, Group Theory and Nonlinear Optics (New York) **13** (2) (2009) – in press.

[121] Study of vibrational dynamics of $Fe_{80}B_{10}Si_{10}$ bulk metallic glass, ***Aditya M. Vora,*** International Journal of Theoretical Physics, Group Theory and Nonlinear Optics (New York) **13** (3-4) (2009) – in press.

[122] Study of vibrational dynamics of $Fe_{40}Ni_{40}B_{20}$ bulk metallic glass, ***Aditya M. Vora,*** International Journal of Theoretical Physics, Group Theory and Nonlinear Optics (New York) **13** (3-4) (2009) – in press.

[123] Study of vibrational dynamics of $Ni_{80}B_{10}Si_{20}$ bulk metallic glass, ***Aditya M. Vora,*** International Journal of Theoretical Physics, Group Theory and Nonlinear Optics (New York) **13** (3-4) (2009) – in press.

[124] Study of vibrational dynamics of $Fe_{60}Ni_{20}B_{10}Si_{10}$ bulk metallic glass, ***Aditya M. Vora,*** International Journal of Theoretical Physics, Group Theory and Nonlinear Optics (New York) **13** (3-4) (2009) – in press.

[125] Computation of Phonon Dispersion Curves of $Ca_{70}Mg_{30}$ Metallic Glass, ***Aditya M. Vora,*** International Journal of Theoretical Physics, Group Theory and Nonlinear Optics (New York) **13** (3-4) (2009) – in press.

[126] Phonon dispersion in Zr-Ni bulk metallic glasses, ***Aditya M. Vora,*** Chemical Physics Research Journal (New York) **2** (3-4) (2009) – in press.

[127] Liquid Metals: A Brief Review, ***Aditya M. Vora,*** Materials Science Research Journal (New York) **3** (1-2) (2009) 1-6.

[128] Stacking faults in the single crystals, ***A. M. Vora and*** M. M. Vora, Journal of Science (Saurashtra University) (India) **1** (1) (2010) 75-77.

[129] Superconducting state parameters of $Nb_xTa_yMo_z$ superconductors, ***Aditya M. Vora,*** Chinese Physics Letters (UK) **27** (2) (2010) 026102 (1)-026102(4).

[130]Study of thermodynamics of liquid noble-metals alloys through a pseudopotential theory, ***Aditya M. Vora,*** Romanian Journal of Physics (Romania) **55** (1-2) (2010) 137-151.

[131] Superconducting state parameters of binary alloys, ***Aditya M. Vora,*** Physica C: Superconductivity and its Applications (Netherlands) **470** (2010) 475-481.

[132] Electron dispersion in liquid alkali and their alloys, ***Aditya M. Vora,*** Communications in Theoretical Physics (China) **54** (1) (2010) 159-166.

[133] Phonon dispersion curves of liquid metals, ***Aditya M. Vora,*** Journal of Non-Oxide Glasses (Romania) **2** (2) (2010) 91-96.

[134] **Phonon dynamics of Fe-based glassy alloys,** ***Aditya M. Vora,*** Chemical Physics Research Journal (New York) **3** (2-3) (2010) 151-170.

[135] Vibrational dynamics of Zr-based bulk metallic glasses, ***Aditya M. Vora,*** Chemical Physics Research Journal (New York) **3** (2-3) (2010) 171-198.

[136] Pressure effects on $Ca_{60}Al_{40}$ metallic glass superconductors, ***Aditya M. Vora,*** Iranian Journal of Physics Research (Iran) **10** (1) (2010) 15-20.

[137] Static and vibrational properties of equiatomic Rb-based binary alloys, ***Aditya M. Vora,*** African Physical Review (Africa) **3** (0003) (2010) 13-19.

[138] Static and vibrational properties of equiatomic binary alloys, Aditya M. Vora, Armenian Journal of Physics (Armenian Academy of Science) (Armenia) **3** (2) (2010) 116-130.

[139] Study of thermodynamical properties of liquid binary alloys by a pseudopotential method, ***Aditya M. Vora,*** Journal of Theoretical and Applied Physics (Iran) **3** (4) (2010) 25-32.

[140] Study of superconducting state parameters of Cu-Zr metallic glasses, ***Aditya M. Vora,*** Electronic Journal of Theoretical Physics (2010) – in press (Online Publications).

[141] Pseudopotential in the study of superconducting state parameters of binary metallic glasses, ***Aditya M. Vora***, International Journal of Modern Physics B (Singapore) (2010) – in press.

[142] Superconducting state parameters of 4d-transition metals based binary alloys, ***Aditya M. Vora,*** Modern Physics Letters B (Singapore) (2010) – in press.

[143] Thermodynamic properties of some liquid metals using pseudopotential theory, ***Aditya M. Vora,*** Physics and Chemistry of Liquids (UK) (2010) – in press.

[144] Study of monovacancy resistivity of some dilute alloys by a pseudopotential theory, ***Aditya M. Vora,*** Materials Physics and Mechanics (Russia) (2010) – in press.

[145] Study of Thermodynamics of liquid noble-metals alloys through a pseudopotential theory, ***Aditya M. Vora,*** Communications in Theoretical Physics (China) (2010) – in press.

[146] Pressure dependence of the superconducting state parameters of binary metallic glass superconductor $Mg_{70}Zn_{30}$, Aditya M. Vora, International Journal of Modern Physics B (Singapore) (2010) – in press.

[147] Study of pressure dependence superconducting state parameters of binary metallic glasses by pseudopotential theory, ***Aditya M. Vora,*** International Journal of Theoretical Physics, Group Theory and Nonlinear Optics (New York) (2010) – in press.

[148] Lattice dynamics of equiatomic alkali binary alloys, ***Aditya M. Vora,*** Materials Science Research Journal (New York) **4**(3) (2010) – in press.

[149] Lattice dynamics of liquid alloys, ***Aditya M. Vora,*** Materials Science Research Journal (New York) **4**(3) (2010) – in press.

[150] Lattice dynamics of solid metallic elements, ***Aditya M. Vora,*** Materials Science Research Journal (New York) **4**(3) (2010) – in press.

[151] Lattice dynamics of liquid metals, ***Aditya M. Vora,*** Materials Science Research Journal (New York) **4**(3) (2010) – in press.

[152] Study of electrical transport properties of liquid semiconductor binary alloys using pseudopotential theory, ***Aditya M. Vora,*** Physics and Chemistry of Liquids (UK) (2010) – in press.

[153] Superconducting state parameters of ternary metallic glasses, ***Aditya M. Vora,*** Bulletin of Materials Science (India) (2010) – in press.

[154] Linearization of the transition temperature equation, ***Aditya M. Vora,*** Journal of Sci-Tech (India) (2010) – in press.

List of Book's Chapter / Review Articles written

[1] Vibrational dynamics of bulk metallic glasses studied by pseudopotential theory, ***Aditya M. Vora,*** In: Computational Materials (Chapter-3), Ed. Wilhelm U. Oster, Nova Science Publishers, Inc., New York (2009) 119-176.

[2] Lattice dynamics of metallic elements and their alloys Studied by Pseudopotential Theory, ***Aditya M. Vora,*** In: Horizons in World Physics, Ed. Albert Reimer, Vol. 270, Nova Science Publishers, Inc., New York (2010)-in press.

[3] Research on Theoretical Physics: A brief introduction, ***Aditya M. Vora,*** In: Horizons in World Physics, Ed. Albert Reimer, Vol. 270, Nova Science Publishers, Inc., New York (2010)-in press.
[4] Study of superconductivity in metallic elements, alloys and metallic glasses using pseudopotential theory, ***Aditya M. Vora,*** In: Advances in Materials Science Research, Ed. Maryann C. Wythers, Vol. 3, Nova Science Publishers, Inc., New York (2010)-in press.
[5] Theoretical studies of thermoelectric power of metallic complexes, ***Aditya M. Vora,*** In: Thermoelectric Power, Ed. William P. Dempsey, Nova Science Publishers, Inc., New York-to be communicated.
[6] Energy research updates: A brief introduction, ***Aditya M. Vora,*** In: Advances in Energy Research, Ed. Morena J. Acosta, Vol. 2, Nova Science Publishers, Inc., New York (2010)-in press.
[7] Theoretical study of superconducting phase in metallic complexes, ***Aditya M. Vora,*** In: Theory and Applications of Superconductivity, Ed. Frank Columbus, Nova Science Publishers, Inc., New York (2011)-in press.
[8] Phonon dispersion in Zr-Ni bulk metallic glasses, ***Aditya M. Vora,*** In: Chemical Physics Research Developments, Ed. Peter W. Hansen, Nova Science Publishers, Inc., New York (2011)-in press.
[9] Phonon dynamics of Fe-based glassy alloys, ***Aditya M. Vora,*** In: New Trends in Chemical Physics Research, Ed. James N. Lapherty, Nova Science Publishers, Inc., New York (2011)-in press.
[10] Vibrational dynamics of Zr-based bulk metallic glasses, ***Aditya M. Vora,*** In: New Trends in Chemical Physics Research, Ed. James N. Lapherty, Nova Science Publishers, Inc., New York (2011)-in press.

LIST OF BOOKLETS WRITTEN

[1] *Some Famous Indian Scientists*, Dr. Aditya M. Vora and Mihir M. Vora, Akshat Publications.
[2] *Some Famous Scientists of the World*, Dr. Aditya M. Vora and Mihir M. Vora, Akshat Publications.
[3] *Grahan na Rahasyo*, Dr. Aditya M. Vora and Mihir M. Vora, Akshat Publications.
[4] *A Brief History of Nanotechnology*, Dr. Aditya M. Vora and Mihir M. Vora, Akshat Publications.
[5] *The Sun – A Brief History*, Dr. Aditya M. Vora and Mihir M. Vora, Akshat Publications.
[6] *The Universe – A Brief History*, Dr. Aditya M. Vora and Mihir M. Vora, Akshat Publications.
[7] *Vigyan na Prayogo*, Dr. Aditya M. Vora and Mihir M. Vora, Akshat Publications.
[8] *Numerical Integration Methods*, Dr. Aditya M. Vora and Mihir M. Vora, Akshat Publications.
[9] *Metallic Glasses – A Brief Review*, Dr. Aditya M. Vora, Akshat Publications.
[10] *Pseudopotential – A Brief Review*, Dr. Aditya M. Vora, Akshat Publications.

LIST OF CONFERENCES / SEMINARS / SYMPOSIA / WORKSHOP ATTENDED

[1] IPA-IC meeting held on March 08-09, 1998 at Department of Physics, Saurashtra University, Rajkot.
[2] 14th Gujarat Science Congress held on October 10-11, 1998 at Palitana, Bhavnagar.
[3] One day seminar on Exploring Professional and Academic Avenues in Physics held on March 7, 1999 at Department of Physics, Sardar Patel University, Vallabh Vidyanagar.
[4] One day seminar on Industrial Application of Materials (IAM-99) held on March 21, 1999 at Department of Physics, M. S. University, Vadodara.
[5] IPA-AC Symposium on Indian Space Programme for New Millennium and the IPA Inter Chapter Meeting held on January 22-23, 2000 at Physical Research Laboratory, Ahmedabad.
[6] 11th AGM of Materials Research Society of India (Gujarat Chapter) held on February 3-5, 2000 at M. S. University of Baroda, Vadodara.
[7] 15th Gujarat Science Congress held on February 12-13, 2000 at Faculty of Science, M. S. University of Baroda, Vadodara.
[8] One day Seminar on Condensed Matter Physics held on March 5, 2000 at Department of Physics, Sardar Patel University, Vallabh Vidyanagar.
[9] 2nd International conference on Different Disordered Systems (INDIAS 2000) held on December 01-03, 2000 at Allahabad.
[10] 43rd DAE Solid State Physics Symposium held on December 27-31, 2000 at Guru Ghasidas University, Bilaspur.
[11] Workshop cum training course on Meteorites, Asteroids and Planets held on February 26-March 02, 2001 at Physical Research Laboratory, Ahmedabad.
[12] One day Seminar on Condensed Matter Physics held on March 17, 2001 at Department of Physics, Sardar Patel University, Vallabh Vidyanagar.
[13] One day Seminar on Condensed Matter Physics held on March 25, 2001 at Department of Physics, M. S. University of Baroda, Vadodara.
[14] International symposium on advances in Superconductivity & Magnetism: Materials, Mechanisms & Devises (ASMM2D), held on 25-28 September, 2001 at Mangalore University, Mangalore.
[15] Gujarat State Research Students Meet-2001 (GSRSM) held on December 22, 2001 at Department of Materials Science, Sardar Patel University, Vallabh Vidyanagar.
[16] One day Seminar on "Experimental Condensed Matter Physics" held on March 23, 2002 at Department of Physics, Sardar Patel University, Vallabh Vidyanagar.
[17] 16th Gujarat Science Congress held on September 14-15, 2002 at Physical Research Laboratory, Ahmedabad.
[18] 4th National Symposium of Research Scholars (NSRS-4) held on September 27-28, 2002 at Indian Institute of Technology Madras (IITM), Chennai.
[19] 2nd Gujarat State Research Students Meet-2003 (GSRSM) held on February 26, 2003 at Sardar Patel University, Vallabh Vidyanagar.
[20] Seminar on Theoretical Physics for Condensed Matter, Atoms-Particles & Particles held on March 21-22, 2003 at Department of Physics, Sardar Patel University, Vallabh Vidyanagar.
[21] Symposium on Condensed Matter Physics (CMDays-2003) held on August 27-29, 2003 at Department of Physics, Jadavpur University, Kolkata.

[22] 73rd Annual Session of The National Academy of Sciences (India) held on October 10-12, 2003 at Physical Research Laboratory (PRL), Ahmedabad.
[23] Indian Physics Association Seminar on "Physics Education, Basic Research and New Technologies" (IAPS–2003) held on November 06-08, 2003 at Bhabha Atomic Research Centre (BARC), Mumbai.
[24] 3rd Gujarat State Research Students Meet-2003 (GSRSM), held on December 09, 2003 at Sardar Patel University, Vallabh Vidyanagar.
[25] 46th DAE Solid State Physics Symposium held on December 26-30, 2003 at Jiwaji University, Gwalior.
[26] Seminar on Condensed Matter Physics held on February 21, 2004 at Department of Physics, Sardar Patel University, Vallabh Vidyanagar.
[27] IPA Inter Chapter Meeting held on March 28, 2004 at Department of Physics, Sardar Patel University, Vallabh Vidyanagar.
[28] The Abdus Salam International Centre for Theoretical Physics (ICTP), Trieste sponsored International School and Conference on "Unifying Concept of Glass Physics III" held on June 25-July 1, 2004 at Jawaharlal Nehru Centre for Advanced Scientific Research (JNCASR), Bangalore.
[29] IAPT-INFOSYS Workshop for UG Teachers (Western Region) held on December 06-11, 2004 at Department of Physics, Sardar Patel University, Vallabh Vidyanagar.
[30] 19th Gujarat Science Congress held on February 19, 2005 at Sardar Patel University, Vallabh Vidyanagar.
[31] Workshop on "Space and GIS Technology Applications" held on February 26, 2005 at Department of Physics, Sardar Patel University, Vallabh Vidyanagar.
[32] National Seminar on "Green Chemistry" held on February 06-08, 2009 at Department of Chemistry, Veer Narmad South Gujarat University, Surat.
[33] 23rd Gujarat Science Congress held on February 15, 2009 at Veer Narmad South Gujarat University, Surat.

LIST OF AICTE APPROVED PROGRAM ATTENDED

[1] AICTE sponsored staff development programme on "Recent Scientific and Technological Advances in Physics Sciences (RSTAPS'08-09)" held on December 29, 2008-January 02, 2009 at Department of Applied Sciences and Humanities, Sardar Vallabhbhai National Institute of Technology (SVNIT), Surat.
[2] AICTE sponsored training programme on "General Research Methodology" held on January 19-23, 2009 at NITTTR Extension Centre, Pune.
[3] AICTE sponsored staff development programme on "Advance Topics in Applied Physics (ATAP-2010)" held on February 01-5, 2010 at Department of Applied Sciences and Humanities, Sardar Vallabhbhai National Institute of Technology (SVNIT), Surat.

Details of Papers Presented at International Level Conferences

[1] P. N. Gajjar, ***A. M. Vora***, M. H. Patel and A. R. Jani, *Some static and vibrational properties of equiatomic sodium based binary alloys*, presented at 2nd International conference on Different Disordered Systems (INDIAS 2000) held on December 01-03, 2000 at Allahabad.

[2] ***A. M. Vora***, M. H. Patel, P. N. Gajjar and A. R. Jani, *Superconducting state parameters of Indium based binary alloys*, presented at International symposium on advances in Superconductivity & Magnetism: Materials, Mechanisms & Devises (ASMM2D), held on 25-28 September, 2001 at Mangalore University, Mangalore.

[3] P. N. Gajjar, ***A. M. Vora*** and A. R. Jani, *Vibrational dynamics of $Ca_{70}Mg_{30}$ metallic glass*, presented as an invited talk in 9th Asia Pacific Physics Conference (APPC) Hanoi, Vietnam, October 25-31, 2004.

Details of Papers Presented at National Level Conferences

[1] ***A. M. Vora***, M. H. Patel, B. Y. Thakore, P. N. Gajjar and A. R. Jani, *Residual resistivity of some simple metals and their alloys*, presented at 11th AGM of Materials Research Society of India (Gujarat Chapter) held on February 3-5, 2000 at M. S. University of Baroda, Vadodara.

[2] ***A. M. Vora***, M. H. Patel, B. Y. Thakore, P. N. Gajjar and A. R. Jani, *Monovacancy resistivity of some simple metals and their alloys*, presented at 43rd DAE Solid State Physics Symposium held on December 27-31, 2000 at Guru Ghasidas University, Bilaspur.

[3] ***A. M. Vora***, M. H. Patel, S. R. Mishra, P. N. Gajjar and A. R. Jani, *Superconducting state parameters of some metallic glasses*, presented at 44th DAE Solid State Physics Symposium held on December 26-30, 2001 at Bhabha Atomic Research Centre (BARC), Mumbai.

[4] ***A. M. Vora***, M. H. Patel, B. Y. Thakore, P. N. Gajjar and A. R. Jani, *Electrical resistivity of Na-based alkali binary alloys*, presented at 45th DAE Solid State Physics Symposium held on December 26-30, 2002 at Panjab University, Chandigarh.

[5] ***A. M. Vora***, P. N. Gajjar and A. R. Jani, *Superconducting state parameters of metals*, presented at 4th National Symposium of Research Scholars (NSRS-4) held on September 27-28, 2002 at Indian Institute of Technology Madras (IITM), Chennai.

[6] P. N. Gajjar, ***A. M. Vora*** and A. R. Jani, *Superconducting state parameters of metals*, presented at Symposium on Condensed Matter Physics (CMDays-2003) held on August 27-29, 2003 at Department of Physics, Jadavpur University, Kolkata.

[7] ***A. M. Vora***, M. H. Patel, P. N. Gajjar and A. R. Jani, *Vibrational dynamics of $Mg_{70}Zn_{30}$ metallic glass*, presented at 46th DAE Solid State Physics Symposium held on December 26-30, 2003 at Jiwaji University, Gwalior.

[8] ***A. M. Vora***, *Study of vibrational dynamics of binary metallic glasses using pseudopotential theory*, Thesis presented at 49th DAE Solid State Physics Symposium held on December 26-30, 2004 at Guru Nanak Dev (GND) University, Amritsar.

[9] ***Aditya M. Vora***, *Effect of indium intercalation on various properties of $MoSe_2$ single crystals*, presented at National Seminar on "Green Chemistry" held on February 06-08, 2009 at Department of Chemistry, Veer Narmad South Gujarat University, Surat.

DETAILS OF PAPERS PRESENTED AT STATE LEVEL CONFERENCES

[1] ***A. M. Vora***, M. P. Deshpande and P. K. Garg, *Effect of Indium intercalation on various properties of $MoSe_2$ single crystals*, presented at One day seminar on Industrial Application of Materials (IAM-99) held on March 21, 1999 at Department of Physics, M. S. University, Vadodara.

[2] ***A. M. Vora*** and P. N. Gajjar, *Monovacancy resistivity of some dilute alloys by pseudopotential theory*, presented at 15th Gujarat Science Congress held on February 12-13, 2000 at Faculty of Science, M. S. University of Baroda, Vadodara.

[3] ***A. M. Vora***, M. H. Patel, B. Y. Thakore, P. N. Gajjar and A. R. Jani, *Residual resistivity of some simple metals and their alloys*, presented at One day Seminar on Condensed Matter Physics held on March 5, 2000 at Department of Physics, Sardar Patel University, Vallabh Vidyanagar.

[4] J. M. Panchal, M. H. Patel, ***A. M. Vora***, P. N. Gajjar and A. R. Jani, *Phonon dispersion for Na-Li, Na-K, Na-Rb and Na-Cs equiatomic solid solutions*, presented at One day Seminar on Condensed Matter Physics held on March 17, 2001 at Department of Physics, Sardar Patel University, Vallabh Vidyanagar.

[5] J. M. Panchal, M. H. Patel, ***A. M. Vora***, P. N. Gajjar and A. R. Jani, *Phonon dispersion for Na-Li, Na-K, Na-Rb and Na-Cs equiatomic solid solutions*, presented at One day Seminar on Condensed Matter Physics held on March 25, 2001 at Department of Physics, M. S. University of Baroda, Vadodara.

[6] ***A. M. Vora***, *Superconducting state parameters of some metals by pseudopotential theory*, presented at 2nd Gujarat State Research Students Meet-2003 (GSRSM), held on February 26, 2003 at Sardar Patel University, Vallabh Vidyanagar.

[7] ***A. M. Vora***, P. N. Gajjar and A. R. Jani, *Electrical resistivity of K-Rb and K-Cs liquid binaries*, presented at Seminar on Theoretical Physics for Condensed Matter, Atoms-Molecules & Particles held on March 21-22, 2003 at Department of Physics, Sardar Patel University, Vallabh Vidyanagar.

[8] ***Aditya M. Vora***, *Superconducting state parameters of metallic elements*, presented at 23rd Gujarat Science Congress held on February 15, 2009 at Veer Narmad South Gujarat University, Surat.

GENERAL DETAILS

ACADEMIC AFFILIATIONS

(1) Life Member of Indian Physics Association, (VVN/LM/11583).

(2) Life member of International Disordered Systems Associates Society (INDIAS), (104/2000/U-I/MDS/India).

(3) Life member of Indian Society for Technical Education (ISTE), (LM/60402/2008).
(4) Member of Physics Alumina Association, Department of Physics, Sardar Patel University, Vallabh Vidyanagar.
(5) Member of Old B. Ed. Students' Association, Shri S. D. Shethiya College of Education, Mundra - Kachchh.

M. Phil. Dissertation Title : Transport Properties Measurement of Indium Intercalated Molybdenum Diselenide Single Crystal-(Experimental Condensed Matter Physics).

Ph. D. Thesis Title : Study of Vibrational Dynamics of Binary Metallic Glasses using Pseudopotential Theory - (Theoretical Condensed Matter & Computational Physics).

M. A. (Education) Dissertation Title : A study of reactions of students and guardians of Surat city towards private tuition.

COMPUTER KNOWLEDGE

LANGUAGES	FORTRAN – 90, C
OPERATING SYSTEMS	DOS 7.0, WINDOWS 95 / 98 / XP
APPLICATION SOFTWARE	MS-OFFICE – 2000 / XP, PAGE MAKER, GRAPHER, MATLAB, MATHCAD

State Level Award : "Hari Ohm Ashram Prerit Bhaikaka Inter-University Smarak Trust" (State Level) award for the best research paper entitled "Fermi Energy and Fermi Surface of Cs-K, Cs-Rb and Rb-K Binary Systems" in Physics (Published in Physica B: Condensed Matter) for the year 2001-2002.

ACADEMIC ACHIEVEMENTS

(1) M. Phil. Student's representative award on the occasion of Physical Society Activities held during the years 1998-1999 at Department of Physics, Sardar Patel University, Vallabh Vidyanagar.
(2) Prof. M. S. Joshi endowment prize for the best research work in physics in the Department of Physics, S. P. University, Vallabh Vidyanagar for the year 2004-2005.
(3) Best cultural activity and student award (for GS) in B. Ed. Course at S. D. Shethiya College of Education, Mundra-Kachchh for the year 2005-2006.

UNIVERSITY RANK: I have secured 2nd rank in B. Ed. course for the year 2005-2006 in KSKV Kachchh University, Bhuj.

TOTAL RESEARCH EXPERIENCES: Total 13.3 years research experience.

TOTAL TEACHING EXPERIENCES: Total 11.1 years teaching experience.

RESEARCH AND TEACHING EXPERIENCES

Post	From	To	Total Experience			Place	Nature of Work / Subject
			Y	M	D		
Project Fellow	01/01/99	30/10/01	02	10	–	Department of Physics, Sardar Patel University, Vallabh Vidyanagar	Research, administrative and teaching work related to UGC Project
Project Fellow	01/12/03	31/03/04	–	04	–	Department of Physics, Sardar Patel University, Vallabh Vidyanagar	Research, administrative and teaching work related to UGC Project
Visiting Lecturer	10/11/01	28/02/05	03	03	18	K. M. Patel Institute of Physiotherapy, P. S. Medical College, Karamsad	Teaching (Bio-Medical Physics)
Adhoc Lecturer	01/09/04	30/09/04	–	01	–	M. B. Patel Science College, Anand	Teaching (Physics)
Lecturer & HOD	07/10/04	31/03/05	–	05	24	Anand Mercantile College of Science, Management and Computer Technology, Anand	Teaching (Physics)
Principal	01/04/05	02/08/08	03	03	03	Sadhu Vaswani International School, Gandhidham - Kachchh	Administration and Teaching (Science / Maths)
Visiting Lecturer	16/07/08	01/08/08	–	–	17	Shri R. R. Lalan College, Bhuj-Kutch	Teaching (Physics)
Lecturer and I/C HOD	04/08/08	Till date	–	–	–	Humanities and Social Science Department, STBS College of Diploma Engineering, Surat	Teaching (Physics, Chemistry, HRM, NCES)

ADMINISTRATIVE / GENERAL EXPERIENCES: I have rendered active services on the following committees/capacity.

(1) Currently, I am working as a I/C Head of Humanities and Social Science Department, STBS College of Diploma Engineering, Surat.

(2) Head and member of various committees of STBS College of Diploma Engineering, Surat such as discipline, anti-ragging, furniture etc.
(3) Nodal Officer of ACPDC help centre, STBS College of Diploma Engineering, Surat.
(4) Recognized Examiner of Gujarat Technological University (GTU), Ahmedabad.
(5) Recognized Examiner of Technical Examination Board (TEB), Gandhinagar.
(6) Former Principal, Sadhu Vaswani International School (English Medium School from K. G. to Higher Secondary), Gandhidham - Kachchh.
(7) Former Head, Physics Department, Anand Mercantile College of Science, Management and Computer Technology, Anand.
(8) Former General Secretary (GS), Shri S. D. Shethiya College of Education, Mundra-Kachchh during 2005-2006.
(9) More than nine years teaching, research and administrative experiences at school, college and university level.
(10) Student member, National Assessment and Accreditation Council (NAAC-UGC), Sardar Patel University-2001.
(11) Project Fellow under UGC sponsored major research projects.
(12) Member of organizing committee, 2nd Gujarat State Research Students Meet-2003.
(13) Member of organizing committee, 3rd Gujarat State Research Students Meet-2003.
(14) Member of organizing committee, Seminar on Condensed Matter Physics-2004.
(15) Member of various local committees in the AMCOST College.
(16) Member of research students committee in the Department of Physics, Sardar Patel University.
(17) Student member of various activities in the Department of Physics, Sardar Patel University.
(18) M. Phil. student's representative in the Department of Physics, Sardar Patel University during 1998-1999.

All India Radio Talk: "Aam Kem? Rojbarojna Jivanma Ghatati Ghatanaoni Vaignanik Samjuti" is broadcasted on 10/04/2008.

Educational Talks: (1) I have given several educational talks at various schools.
(2) I have given pedagogic talk on "General Research Methodology" in ISTE faculty chapter program at STBS College of Diploma Engineering, Surat on 06/03/2010.

Articles Publications (1) Global Warming – Aak Jatil Samasya – Published in "Aajkaal-Daily" on 13/04/2008.
(2) Navi Pedhine Paymaal Kartu Sisanu Pradushan – Published in "Aajkaal-Daily" on 20/04/2008.
(3) Education and Socialization-Published in College Magazine of STBS College of Diploma Engineering, Surat-"V-will" (2009), p. 16.

Extra Curricular Activities: Sports, Cultural Activities, Reading, G.K. etc..

I hereby declare that all the information provided above is true to the best of my knowledge.

Date: 30/09/10 (Dr. Aditya M. Vora)
Place: Surat

In: Superconductivity and Condensed Matter ...
Editor: Collin J. Eckles
ISBN 978-1-61209-785-5

Chapter 31

BIOGRAPHICAL SKETCH – Superconductivity Research

NAME:	TITLE:	DATE OF BIRTH:
KUNIHIKO YAMAJI	DR. AND PROFESSOR	APRIL 18, 1942

EDUCATION

Institution and Location	Degree	Year Conferred	Scientific Field
University of Tokyo, Tokyo University of Tokyo, Tokyo	Baccalaureate Master Doctor of Sciences	1965 1967 1971	Physics Theory of solid-state physics Theory of solid-state physics

CONTACT POINTS

Email: yamaji-kuni@aist.go.jp Phone: 81-298-62-1997 Mail address: AIST Central2, 1-1-1 Umezono, Tsukuba 305-8568, Japan

RESEARCH AND PROFESSIONAL EXPERIENCE

April 2003 – Present Technical Staff (contract employee), Nanotechnology Research Institute (NRI), National

Advanced Institute of Industrial Science and Technology (AIST), Japan May 2004 – Nov. 2006 Board Member of “Report on Progress in Physics”, Institute of Physics, UK April 2003 – March 2004 Part-time Lecturer, Physics Department, Toho University, Japan March 2003 Retired from AIST as a regular member due to age of 60 April 2001 Principal Researcher, NRI, AIST, Japan (I moved to NRI due to restructuring of

Electrotechnical Laboratory (ETL) inside AIST) April 1995 – March 1998 Adjunct Professor, Physics Department, Osaka University, Japan April 1992 – March 2002 Adjunct

Professor, Materials Science Institute, Tsukuba University, Japan Jan. 1994 Chief Senior Researcher, ETL April 1990 Chief, Fundamentals of Solid-State Physics Section, ETL Jan. – April 1987 Visiting Professor, Solid-State Physics Laboratory, Universite Paris-Sud, France April 1979 – Aug. 1980 Visiting Associate Professor, the above-mentioned institute July 1974 Senior Researcher, ETL April 1971 Researcher, Low Temperature Physics Section, Electrotechnical Laboratory (ETL), Japan

Jan. 2004 -Fellow, Institute of Physics, UK April 1996 Article Award of Physical Society of Japan May 1994 General-Director's Commendation, AIST, Japan Aug. 1986 Research Achievement Award, ETL, AIST, Japan

Professional Appointments: Honors Relevant to Superconductivity Research

Publications During Last Three Years

K. Yamaji, T. Yanagisawa, I. Hase, Physica C, In Press (2010) "3-Band Theory Of Fe Pnictide Superconductors"

K. Yamaji, T. Yanagisawa, M. Miyazaki, R. Kadono, Physica C, 469 (2009) 1037, "System-Parameter Dependence Of The Metallic Phase Of The Non-Doped 2d Hubbard Model"

T. Yanagisawa, K. Odagiri, I. Hase, K. Yamaji, P. M. Shirage, Y. Tanaka, A. Iyo, H. Eisaki., J. Phys. Soc. Japan 78 (2009) 094718, "Isotope Effect In Multi-Band And Multi-Channel Attractive Systems And Inverse Isotope Effect In Iron-Based Superconductors"

M. Miyazaki, K. Yamaji, T. Yanagisawa, R. Kadono, J. Phys. Soc. Japan 78 (2009) 043706, "Checkerboard States In The Two-Dimensional Hubbard Model With The Bi2212-Type Band"

T. Yanagisawa, M. Miyazaki, K. Yamaji, J. Phys. Soc. Japan 78 (2009) 13706, "Incommensurate Antiferromagnetism Coexisting With Superconductivity In Two-Dimensional D-P Model"

M. Miyazaki, K. Yamaji, T. Yanagisawa, R. Kadono, Physica C, In Press (2009) "Coexistence Between Checkerboarded Charge And Striped Spin In The Two-Dimensional Hubbard Model"

K. Yamaji, T. Yanagisawa, M. Miyazaki, R. Kadono, Physica C, 468 (2008) 1125, "T'- And T"-Dependence Of The Bulk-Limit Superconducting Condensation Energy Of The 2d Hubbard Model"

In: Superconductivity and Condensed Matter ...
Editor: Collin J. Eckles

ISBN 978-1-61209-785-5

Chapter 32

BIOGRAPHICAL SKETCH – Superconductivity Research

NAME:	TITLE:	DATE OF BIRTH:
ARCADY ZHUKOV	DR. IN PHYSICS	28.12.1956

EDUCATION

Institution and Location	Degree	Year Conferred	Scientific Field
Physics-Chemistry Department of Moscow Steel and Alloys Institute (presently Moscow Technological University)	Engineer	1980	Material Science
Institute of Solid State Physics of Russian Academy of Sciences	PhD	1988	Solid State Physics
Institute of Apllied Magnetism, Computense University, Madrid, Spain (Spanish Ministry of Education fellowship)	Postdoctoral position	1994-1996	Magnetic Materials

CONTACT POINTS

Department Of Physics of Materials, Chem. Faculty, University of the Basque Country, P. Manuel de Lardizabal, 3, 20018 Spain

Research and Professional Experience

Professional Appointments

Employment record:
Physics-Chemistry Department of Moscow Steel and Alloys Institute (presently Moscow Technological University)

Student: 1974-1980
(Head of Department and professor of the "Solid State Physics" – Prof. Alexei Abrikosov, Nobel Price in Physics, 2003)

Institute of Solid State Physics	research probationer	1980-1982
of Russian Academy of Sciences (ISSP)	post-graduate	1982-1986
	research associate	1986-1994
	Ph.D.	1988
	supervisor - Prof. B.K.Ponomarev	
Instituto de Magnetismo Aplicado, Madrid, Spain,	postdoctoral fellowship	1994-1996
European Community, Programme ESPRIT	fellowship	February -December 1996
European Community, Programme MUSIC	fellowship	January -June 1997
University of Basque Country - San Sebastian, Spain -	visiting scientist	July 1997- June 1999
University of Basque Country - San Sebastian, Spain -	contracted scientist	July 1999- June 2000
Donostian International Physics Centre, University of Basque Country	invited professor	July 2000 - December 2001
Institute of Material Science, CSIC - (Council Superior of Research, Spain)	researcher (Ramon y Cajal prog.)	January 2002 - October 2003
University of Basque Country - San Sebastian, Spain	researcher	November 2003 up to now

Selected for the Ikerbasque Research Professor (July 2010)

HONORS RELEVANT TO RESEARCH:

Awarding by foundations

Individual grant of the International Science Foundation, 1993
Premio de Manuel Laborde Werlinden, 2004

PUBLICATIONS DURING LAST THREE YEARS

BOOKS

1. A. Zhukov, J. M. Blanco, J. González and V. Zhukova, Chapter "Magnetic Properties Of Glass-Coated Microwires With Nanocrystalline Structure", Book: "Magnetic Materials: Research, Technology", Isbn: 978-1-60692-145-6 Editor: Jacob I. Levine, © 2009 Nova Science Publishers, Inc., Nova Science Publishers, Inc. 400 Oser Avenue, Suite 1600 Hauppauge, NY 11788
2. A. Zhukov and V. Zhukova, «Magnetic properties and applications of ferromagnetic microwires with amorphous and nanocrystalline structure", Nova Science Publishers, Inc. 400 Oser Avenue, Suite 1600 Hauppauge, NY 11788, 162 p. 2009, ISBN: 978-1-60741-770-5.
3. Journal Of Nanoscience And Nanotechnology (Special issue devoted to nanomagnetism), Guest Editors O. Chubykalo-Fesenko, J. Gonzalez, B. Hernando, and A. Zhukov, Volume: 8 Issue: 6 Pages: 2729-2729 Published: JUN 2008
4. J. González and A. Zhukov, "Surface and Bulk Magnetism of Materials with Amorphous and Nanostructure Character", *Nanoclusters and Nanostructured Surfaces* Edited byAsok K. Ray Pages: 1–45, chapter 9, 2010, American Scientific Publishers (accepted), ISBN: 1-58883-182-5
5. A. Zhukov, A. Chizhik, J. J. d el Val, J. M. Blanco, V. Zhukova, "Nanocrystalline Structure in Amorphous Magnetic Wires, Glass Covered Microwires and Ribbons", *Nanoclusters and Nanostructured Surfaces* Edited byAsok K. Ray Pages: 1–35, chapter 1, 2010, American Scientific Publishers (accepted), ISBN: 1-58883-182-5
6. *AUTORES (p.o. de firma):* N. Iturriza, J.J. del Val, A. Zhukov, I. García, J. A. Pomposo and J. González
   TITULO: Novel Amorphous and Nanocrystalline Soft Magnetic Materials
   REF. REVISTA/LIBRO: Amorphous Materials: Research, Technology and Applications, Ed. Jacob I. Levine, (NOVA Science Publishers, New York, 2009) Vol.3, Chapter 5, p. 84-139) (ISBN: 978-1-60692-235-4)

SCIENTIFIC JOURNALS

INVITED PAPERS

1. A. Zhukov, M. Ipatov, C. García, J. Gonzalez, J. M. Blanco and V. Zhukova, Magnetic Properties and High–Frequency GMI Effect in Thin Glass-Coated Amorphous Wires , AIP Conf. Proc. 1003 (2008) 280-286
2. D. Makhnovskiy, A. Zhukov, V. Zhukova, J. Gonzalez, "Tunable and self-sensing microwave composite materials incorporating ferromagnetic microwires" *Advances in Science and Technology Vol. 54 (2008) pp 201-210*
3. L V. Panina , D.P. Makhnovskiy, A. Zhukov and J. Gonzalez, "Multilayered Magnetic Wires and Films for Electromagnetic Sensor Technology" *Advances in Science and Technology Vol. 54 (2008) pp 29-40*
4. A.Zhukov, M.Ipatov, J.Gonzalez, J.M.Blanco, V.Zhukova, Recent advances in studies of magnetically soft amorphous microwires, J. Magn. Magn. Mater. 321 (2009) 822–825
5. M. Ipatov, V. Zhukova, L. V. Panina, and A. Zhukov "Ferromagnetic Microwires Composite Metamaterials with Tuneable Microwave Electromagnetic Parameters" (invited paper), Proceedings of Conference Progress In Electromagnetics Research Symposium Proceedings, Moscow, Russia, August 18-21, 2009 pp. 1657-1661
6. V. Zhukova, M. Ipatov and A Zhukov, "Thin Magnetically Soft Wires for Magnetic Microsensors" (*Review*) *Sensors 9*(2009) pp. 9216-9240
7. L. V. Panina, M. Ipatov, V. Zhukova, J. Gonzalez and A. Zhukov, "Microwave Metamaterials With Ferromagnetic Microwires", Proceedings of META'10, International Conference on Metamaterials, Photonic crystals and Plasmonics, Edited by Saïd Zouhdi, University Paris-Sud, France, January 18th, 2010 pp.248-254

REGULAR CONTRIBUTIONS

8. V. Zhukova, A. Zhukov, F. J. Palomares, F. Pigazo, F. Cebollada, J. J. Del Val, C. García, J.M.Gonzalez and J. Gonzalez, "Thermal dependence of coercivity in granular CoNiCu glass coated microwires", J. Magn. Magn. Mater. 310 (2007) e867–e869
9. R. Varga, A. Zhukov, V. Zhukova, J. M. Blanco and J. Gonzalez, "Supersonic domain wall in magnetic microwires", Phys. Rev. B **76**, (2007,)132406
10. V. Zhukova, M. Ipatov, C. García, J. Gonzalez, J. M. Blanco and A. Zhukov, "Development of Ultra-Thin Glass-Coated Amorphous Microwires for High Frequency Magnetic Sensors Applications", *Open Materials Science Reviews, 1,(2007), pp.*1-12
11. M. Vázquez, G. Badini, K. Pirota, J. Torrejón, A. Zhukov, A. Torcunov, H. Pfützner, M. Rohn, A. Merlo, B. Marquardt and T. Meydan, " Applications of amorphous microwires in sensing technologies", International Journal of Applied Electromagnetics and Mechanics, Volume 25, Number 1-4 / 2007, **Pp:** 441 – 446
12. A. Chizhik, C. Garcia, A. Zhukov, P. Gawronski, K. Kulakowski, J. Gonzalez, J.M. Blanco, "Investigation of helical magnetic structure in Co-rich amorphous microwires", J. Magn. Magn. Mater. 316 (2007) pp. 332-336

13. C. Garcia, A. Zhukov, V. Zhukova, V. Larin, J. Gonzalez, J.J. del Val, M. Knobel, "Temperature dependence of magnetic properties of Cu80Co19Ni1 thin microwires"J. Magn. Magn. Mater. 316 (2007) pp. e71-e73
14. C. Garcia, A. Chizhik, A. Zhukov, V. Zhukova,J. Gonzalez, J.M. Blanco, L.V. Panina, "Influence of torsion and tensile stress on magnetoimpedance effect in Fe-rich amorphous microwires at high frequencies", J. Magn. Magn. Mater. 316 (2007) pp. e96-e99
15. A. Chizhik, J.M. Blanco, A. Zhukov, J. Gonzalez, C. Garcia, P. Gawronski, K. Kulakowski, "Magneto-optical determination of helical magnetic structure in amorphous microwires", Physica B 403 (2008) 289–292
16. C. García, V. Zhukova, J. Gonzalez, J.M. Blanco, A. Zhukov, "Effect of magnetic field frequency on coercivity behavior of nanocrystalline Fe79Hf7B12Si2 glass-coated microwires", Physica B 403 (2008) 286–288
17. V. Zhukova, J.M. Blanco, M. Ipatov, R. Varga, J. Gonzalez, A. Zhukov, "Domain wall propagation in Fe-rich microwires" Physica B 403 (2008) 282–285
18. M. Ipatov, N.A. Usov, A. Zhukov, J. Gonzalez, "Local nucleation fields of Fe-rich microwires and their dependence on applied stresses", Physica B 403 (2008) 379–381
19. R. Varga, A. Zhukov, N. Usov, J.M. Blanco, J. Gonzalez, V. Zhukova, P. Vojtanik, "Domain-wall dynamics in glass-coated magnetic microwires", J. Magn. Magn. Mater. 316 (2007) 337–339
20. R. Żuberek, H. Szymczak, M. Gutowski, A. Zhukov, V. Zhukova, N.A. Usov, K. Garcia, M. Vazquez, "Internal stress influence on FMR in amorphous glass-coated microwires", J. Magn. Magn. Mater. 316 (2007) e890-e892.
21. N.A. Usov, A. Zhukov, J. Gonzalez, Domain walls and magnetization reversal process in soft magnetic nanowires and nanotubes, ", J. Magn. Magn. Mater. 316 (2007) pp. 255–261.
22. S. Gudoshnikov, N. Usov, A. Zhukov, J. Gonzalez, P. Palvanov, "Measurements of stray magnetic fields of amorphous microwires using scanning microscope based on superconducting quantum interference device", J. Magn. Magn. Mater. 316 (2007) pp. 188-191.
23. A. Chizhik, J. Gonzalez, P. Gawronski, K. Kulakowski, A. Zhukov, J. M. Blanco, "Torsion and tension stress induced transformation of surface magnetic structure in Co-rich amorphous microwires", Journal of Non-Crystalline Solids 353 (2007) 935–937
24. M. Ipatov, R. Varga, A. Zhukov, J. M. Blanco, J. Gonzalez, V. Zhukova "Complex susceptibility measurements in amorphous glass-coated microwires",Journal of Non-Crystalline Solids 353 (2007) 928–930
25. N. A. Usov, A. Zhukov, J. Gonzalez, "Single-domain particle with random anisotropy",Journal of Non-Crystalline Solids 353 (2007) 796–798
26. A. Zhukov, C. García , V. Zhukova, V. Larin, J. González,J. J. del Val, M. Knobel, J. M. Blanco, "Fabrication and magnetic properties of Cu50(Fe69Si10B16C5)50 thin microwires", Journal of Non-Crystalline Solids 353 (2007) 922–924
27. A. Chizhik, J. M. Blanco, A. Zhukov, J. Gonzalez, C. Garcia, P. Gawronski and K. Kulakowski "Surface and Bulk Magnetic Hysteresis Loops of Co-Rich Glass Covered Microwires", IEEE TRANSACTIONS ON MAGNETICS 42, No12, (2006) pp.3889-3992
28. A. Chizhik, R. Varga A. Zhukov J. Gonzalez and J. M. Blanco, "Kerr-effect based Sixtus-Tonks experiment for measuring the single domain wall dynamics", J. Appl. Phys. 103, (2008) 07E707 -1-3

29. A. Chizhik, C. Garcia, A. Zhukov, J. Gonzalez, P. Gawronski, K. Kulakowski, and J. M. Blanco "Relation between surface magnetization reversal and magnetoimpedance in Co-rich amorphous microwires", J. Appl. Phys. 103, (2008) 07E742 -1-3
30. V. Zhukova, M. Ipatov, J. Gonzalez, J. M. Blanco and A. Zhukov, "Studies of magnetic properties and giant magnetoimpedance effect in ultrathin magnetically soft amorphous microwires", : J. Appl. Phys. 103, (2008) 07E714 -1-3
31. G. R. Aranda, N. A. Usov, V. Zhukova, A. Zhukov, and J. Gonzalez, "Magnetostatic propertiesof Co-rich amorphous microwires: theory and experiment", Phys. stat. sol. (a) 205, No. 8, (2008)1800–1804
32. A. Chizhik, J. Gonzalez, A. Zhukov, J.M. Blanco, Transformation of surface domain structure in Co-rich amorphous wires, Sensors and Actuators B 126 (2007) 235–239
33. V. Zhukova, J.M. Blanco, M. Ipatov, A. Zhukov, C. Garcıa, J. Gonzalez, R. Varga, A. Torcunov, Development of thin microwires with low Curie temperature for temperature sensors applications, Sensors and Actuators B 126 (2007) 318–323
34. V. Zhukova, M. Ipatov, A. Zhukov, J. Gonzalez, J.M. Blanco, GMI effect in ultra-thin glass-coated Co-rich amorphous wires, Sensors and Actuators B 126 (2007) 232–234
35. K.Richter, Y. Kostyk, R. Varga, A. Zhukov, and V. Larin , Domain wall dynamics in amorphous microwires, ACTA PHYSICA POLONICA A Volume: 113 Issue: 1 Pages: 7-10 Published: JAN 2008
36. J. Olivera, J.Gamcova, R. Varga, J.D.Santos, V.M.Prida, M.L.Sanchez, B. Hernando, A.Zhukov, Susceptibility spectroscopy in FeNiSiB microwires, Volume: 113 Issue: 1 Pages: 155-158 Published: JAN 2008
37. A. Zhukov, M. Ipatov, V. Zhukova, C. García, J. Gonzalez, and J. M. Blanco, " Development of ultra-thinglass-coated amorphous microwires for HF magnetic sensor applications", Phys. Stat. Sol. (a), 205 No 6 (2008) 1367-1372
38. M.N. Baibich, G. Martínez, M.G.M. Miranda, A.T. da Rosa, J. González, A. Zhukov, Ribbons and micro-wires of CuCo segregated alloys, J. Magn. Magn. Mater. 320 (2008), e29-e31
39. : M. I. Ilyn, V. Zhukova, J. D. Santos, M. L. Sánchez, V. M. Prida, B. Hernando, V. Larin, J. González, A. M. Tishin, and A. Zhukov, "Magnetocaloric effect in nanogranular glass coated microwires" ,Phys. Stat. Sol. (a), 205 No 6 (2008) 1378-1381
40. N. Iturriza, A. Zhukov, J.J. del Val, N.Murillo, J. Gonzalez,, Microstructure and soft magnetic properties of nanocrystalline (C0.77Si0.135B0.095)(90)Fe7Nb3 alloy, Phys. Stat. Sol. (a), 205 (2008) 1363-1366
41. N.A. Usov, A. Zhukov, J. Gonzalez, Title: Remanent magnetization states in soft magnetic nanowires, IEEE TRANSACTIONS ON MAGNETICS Volume: 42 Pages: 3063-3065 Published: 2006
42. M. Ipatov, V. Zhukova, J. M. Blanco, J. Gonzalez, and A. Zhukov, Off-diagonal magneto-impedance in amorphous microwires with diameter 6–10 μm and application to linear magnetic sensors, Phys. stat. sol. (a) 205, No. 8, (2008) 1779–1782
43. A. Zhukov, V. Zhukova, J. Gonzalez, L. Panina and J.M. Blanco, "Development of Stress and Temperature Sensitive Microwires for the Sensor Applications and Tuneable Composite Materials", *Advances in Science and Technology Vol. 54 (2008) pp 180-186*

44. R. Varga, Yu. Kostyk, K. Richter, A. Zhukov, and M. Vazquez, "Domain-wall dynamics in bistable magnetic microwires", Phys. Status Solidi A 206, No. 4, 608–612 (2009)
45. S. A. Gudoshnikov, Yu. B. Grebenshchikov, B. Ya. Ljubimov, P. S. Palvanov, N. A. Usov, M. Ipatov, A. Zhukov, and J. Gonzalez, "Ground state magnetization distribution and characteristic width of head tohead domain wall in Fe-rich amorphous microwire", Phys. Status Solidi A 206, No. 4, 613–617 (2009)
46. V. Popov, V. Zhukova, M. Ipatov, C. García, J. Gonzalez, V. Ponomarenko, V. Berzhansky, D. Vinogorodsky, and A. Zhukov, "Studies of giant magnetoimpedance effect of Co-rich microwires in wide frequency range", P hys. Status Solidi A 206, No. 4, 671–673 (2009)
47. A. Zhukov, M. Ipatov, J. M. Blanco, J. Gonzalez, and V. Zhukova, "Studies of thin microwires with enhanced magnetic softness and GMI effect", Phys. Status Solidi A 206, No. 4, 674–678 (2009)
48. V. Zhukova, J. M. Blanco, M. Ipatov, J. Gonzalez, and A. Zhukov, Domain-wall propagation in thin Fe-rich glass-coated amorphous wires, Phys. Status Solidi A 206, No. 4, 679–682 (2009)
49. A.Chizhik, A.Zhukov, J.M.Blanco, J.Gonzalez, P.Gawronski, "Experimental determination of limit angle of helical anisotropy in amorphous magnetic microwires", J. Magn. Magn. Mat. 321 (2009) 803–805
50. A Zhukov, C Garcıa, J J Del Val, J Gonzalez, M Knobel, D Serantes, D Baldomir and V Zhukova, "Studies of Fe–Cu microwires with nanogranular structure", J. Phys.: Condens. Matter 21 (2009) 035301 (6pp)
51. R. Varga, K. Richter, A. Zhukov, and V. Larin, "Domain Wall Propagation in Thin Magnetic Wires", IEEE TRANSACTIONS ON MAGNETICS Volume: 44 Issue: 11 Pages: 3925-3930 Part: Part 2 Published: NOV 2008
52. V. Zhukova, M. Ipatov, J. González, J. M. Blanco and A. P. Zhukov, " Development of Thin Microwires With Enhanced Magnetic Softness and GMI" , IEEE TRANSACTIONS ON MAGNETICS, VOL. 44, NO. 11, Part: Part 2pages 2958-3961, NOVEMBER 2008
53. J. Olivera, M. L. Sánchez, V. M. Prida, R. Varga, V. Zhukova, A. , and B. Hernando, "Temperature Dependence of the Magnetization Reversal Process and Domain Structure in $Fe_{77.5-x}Ni_xSi_{7.5}B_{15}$ Magnetic Microwires", IEEE TRANSACTIONS ON MAGNETICS Volume: 44 Issue: 11 Pages: 3946-3949 Part: Part 2 Published: NOV 2008
54. A. Chizhik, A. Zhukov, J. M. Blanco, J. Gonzalez, P. Gawronski, and K. Kulakowski, "Experimental Determination of Relation Between Helical Anisotropy andTorsion Stress in Amorphous Magnetic Microwires", IEEE TRANSACTIONS ON MAGNETICS Volume: 44 Issue: 11 Pages: 3938-3941 Part: Part 2 Published: NOV 2008
55. N. A. Usov, A. P. Chen, A. Zhukov, and J. González, "Nucleation field of a soft magnetic nanotube with uniaxial anisotropy", J. APPL. PHYS. 104 (2008) 083902
56. M. Ilyn, A. V. Andreev, V. Zhukova, A. Zhukov, A. Tishin, and J. Gonzalez, "Magnetocaloric effect and spin reorientation transition in single-crystal $Er_2(Co_{0.4}Fe_{0.6})_{17}$", J. APPL. PHYS. 105 (2009) 07A918
57. J. Olivera, R. Varga, P. Vojtanik, V.M. Prida, M.L. Sanchez, B. Hernando, A. Zhukov, "Fast domain wall dynamics in amorphous glass-coated microwires", Journal of Magnetism and Magnetic Materials 320 (2008) 2534–2537
58. V. M. Prida, V. Vega, D. Serantes, D. Baldomir, M. Ilyn, A. Zhukov, J. Gonzalez, and Blanca Hernando, "Influence of magnetic anisotropy and dipolar interactions on

magnetocaloric effect in nanostructured materials", Phys. Status Solidi A, 1–6 (2009) / DOI 10.1002/pssa.200881731

59. A. Chizhik, A. Zhukov, J. Gonzalez, and J. M. Blanco, "Control of domain nucleation in glass covered amorphous microwires", J. APPL. PHYS. 105, (2009) 123911
60. H.X. Peng , F.X. Qin, M.H. Phan, Jie Tang, L.V. Panina, M. Ipatov, V. Zhukova, A. Zhukov, J. Gonzalez, "Co-based magnetic microwire and field-tunable multifunctional macro-composites", Journal of Non-Crystalline Solids 355 (2009) 1380–1386
61. C. García , A. Zhukov, J. González, J.J. del Val, J.M. Blanco, M. Knobel, V. Zhukova" Fabrication, structural and magnetic characterization of thin microwires with novel composition $Cu_{70}(Co_{70}Fe_5Si_{10}B_{15})_{30}$", Journal of Alloys and Compounds 483 (2009) 566–569
62. J.D. Santos, R.Varga, B.Hernando, A.Zhukov "Enhancement of GMI effect in magnetic microwires through the relative temperature dependence of magnetization and anisotropy", Journal of Magnetism and Magnetic Materials 321 (2009) 3875–3877
63. V. S. Larin and A. Zhukov, "Magnetic properties of microwires with amorphous structure after thermo mechanical treatment" Phys. Status Solidi C 6, No. 4, 958–961 (2009)
64. C. García, V. Zhukova, J. J. del Val1, J. M. Blanco, and A. Zhukov, "Studies of electrical resistance in Ni75Cr7Si7.5Mn10.5 and Ni80.5Cr4.2Si6.5Mn5B3.8 glass-coated wires", Phys. Status Solidi C 6, No. 4, 953– 957 (2009)
65. D. Serantes, D. Baldomir, and M. Pereiro, B. Hernando, V. M. Prida, and J. L. Sánchez Llamazares, A. Zhukov, M. Ilyn, and J. González "Magnetocaloric effect in dipolar chains of magnetic nanoparticles with collinear anisotropy axes" Phys. Rev. B 80, 134421 _(2009)
66. M. Ipatov,V. Zhukova, A. K. Zvezdin and A. Zhukov, "Mechanisms of the ultrafast magnetization switching in bistable amorphous microwires", J. Appl. Phys. ,106, 103902, 2009
67. C. García, V. Zhukova, M. Ipatov, J. González, J.M. Blanco, A. Zhukov, "High-frequency GMI effect in glass-coated amorphous wires", Journal of Alloys and Compounds 488 (2010) 9–12
68. C. García, V. Zhukova, J. Gonzazez, A. Chizhik, J.M. Blanco, M. Ipatov, A. Zhukov, "Magnetic and transport properties of Fe-rich thin cold-drawn amorphous wires", Journal of Alloys and Compounds 488 (2010) 5–8
69. V. Zhukova, J. M. Blanco, M. Ipatov and A. Zhukov, "Effect of transverse magnetic field on domain wall propagation inmagnetically bistable glass-coated amorphous microwires", J. Appl. Phys., 106, 113914, 2009
70. *A. Chizhik*, A. Zhukov, A. Stupakiewicz, A. Maziewski, J. Blanco and J. Gonzalez, "Kerr microscopy study of magnetic domain structure changes in amorphous microwires.", IEEE Trans. Magn. VOL. 45, NO. 10, OCTOBER 2009 p. 4279-4281
71. D Serantes, D Baldomir, M Pereiro, B Hernando, V M Prida, J L Sánchez Llamazares, A Zhukov, M Ilyn and J González, "Magnetic ordering in arrays of one-dimensional nanoparticle chains", J. Phys. D: Appl. Phys. 42 (2009) 215003 (9pp)
72. P A Ekstrom and A Zhukov, "Spatial structure of the head-to-head propagating domain wall in glass-covered FeSiB microwire", J. Phys. D: Appl. Phys. 43 (2010) 205001 (6pp)

73. M. Ipatov, V. Zhukova, A. Zhukov, J. Gonzalez, and A. Zvezdin, "Low-field hysteresis in the magnetoimpedance of amorphous microwires", PHYSICAL REVIEW B 81, (2010), 134421.
74. F. X. Qin, N. Pankratov, H. X. Peng, M. H. Phan, L. V. Panina, M. Ipatov, V. Zhukova, A. Zhukov, and J. Gonzalez, "Novel magnetic microwires-embedded composites for structural health monitoring applications", J. APPL. PHYS. 107 (2010) 09A314
75. Olivera J., M. Ipatov, M. L. Sánchez, V. M. Prida, R.Varga, B. Hernando, and A. Zhukov "Pinning Field Distribution and Microstructural Study of Thermal Annealed Fe-Nb-Cu-Si-B Wires", IEEE TRANSACTIONS ON MAGNETICS, VOL. 46, NO. 2, FEBRUARY 2010, 387-389
76. Rodionova V., Ipatov M., Ilyn M., Zhukova V., Perov N., Gonzalez J., and Zhukov A., Design of magnetic properties of arrays of magnetostatically coupled glass-covered magnetic microwires// Phys. Status Solidi A, 1–6 (2010) / DOI 10.1002/pssa.200925497;
77. V. Rodionova, M. Ipatov, M. Ilyn, V. Zhukova, N. Perov, L. Panina, J. Gonzalez and A. Zhukov, "Magnetostatic interaction of glass-coated magnetic microwires" Journal of Applied Physics, 108 (2010) 016103.
78. A. Chizhik, A. Stupakiewicz, A. Maziewski, A. Zhukov, J. Gonzalez and J. M. Blanco, "Direct observation of giant Barkhausen jumps in magnetic microwires", APPLIED PHYSICS LETTERS 97, 012502 (2010)

III. PROCEEDINGS OF INTERNATIONAL CONFERENCES

- With scientific evaluation

79. V. Zhukova, J. M. Blanco, M. Ipatov, J. Gonzalez, and A. Zhukov, Domain Wall Propagation in Thin Fe-Rich Glass-Coated Amorphous Wires , AIP Conf. Proc. 1003 (2008) 301
80. A. Chizhik, A. Zhukov, V. Zhukova, C. Garcia, J. M. Blanco, J. J. del Val, L. Fernandez, N. Iturriza and J. Gonzalez, "Nanocrystallization and Surface Magnetic Structure of Ferromagnetic Ribbons and Microwires" in Advances in Nanoscale Magnetism, ISSN 0930-8989, Springer Berlin Heidelberg, Volume 122, ISBN 978-3-540-69881-4 (Print) 978-3-540-69882-1 (Online), pp. 205-217
81. V V Samsonova, A S Antonov, N A Buznikov, A A Rakhmanov and A. Zhukov , "Experimental study of surface domain structure effects on off-diagonal magnetoimpedance in glass-coated Co-based microwires", Journal of Physics: Conference Series **98** (2008) 062004
82. A. Zhukov, M. Ipatov, C. García, J. Gonzalez, L. Panina, J. M. Blanco and V. Zhukova, "Development of Thin Soft Magnetic Amorphous Microwires for High Frequency Magnetic Sensors Applications", PIERS Proceedings, Hangzhou, China, March 24-28, 2008, pp.650- 657, 2008
83. M. Ipatov, L. V. Panina ,V. Zhukova, J. Gonzalez and A. Zhukov, J.M. Blanco, "Ferromagnetic Microwires Composite Metamaterials with Tunable Microwave Electromagnetic Parameters" REF. REVISTA/LIBRO: COMATCOMP (Materiales compuestos 09- Science and technology of composite materials 09) Proceedings, Editors: A. Güemes, J.Kenny, A. Ureña, A. Vázquez and I. Mondragon, Octubre 2009, San Sebastián, pp.393-396

84. M. Ipatov, G.R. Aranda, V.Zhukova, L. V. Panina, J. González, and A. Zhukov, "Tunable effective permittivity of composites based on ferromagnetic microwires with high magneto-impedance effect", Proceedings of META'10, International Conference on Metamaterials, Photonic crystals and Plasmonics, Edited by Saïd Zouhdi, University Paris-Sud, France, January 18th, 2010 pp.138-144
85. L. V. Panina, M. Ipatov, V. Zhukova, J. Gonzalez, and A. Zhukov, "Tuneable dielectric properties of composites with arrays of magnetic wires ", Proceedings of 3rd International Congress on Advanced Electromagnetic Materials in Microwaves and Optics, London 30 Aug-4Sept, 2009 (ISBN 978-0-9551179-6-1) pp. 788-790, ©2009 Metamorphose-VI

Other contributions of interest to research: Projects, Grants etc.

Projects

Participaton in Research Proyectos

PROJECT: Development of Magnetic Micro-Wires for Technical Applications (DEVMAGMIWIRTEC)
CALL: MANUNET-2007-Basque-3 (ERA-NET)
PARTICIPANTS: Universidad del País Vasco, Tamag Ibérica (Coordinador), Univerity of Pavol Jozef Safarik in Kosice, Company EDIS (Slovak Republic), Faculty of Aeronautics, Technical University in Kosice (Slovak Republic), EAR Aplicaciones Electronicas S.L. (Spain)
From: 01/01/2008 *To*: 31/12/2010
PRINCIPAL Researcher (coordinator): Arcady Zhukov

PROJECT: Artificially structured Materials (metamaterials) based on microwires with high GMI effect in the microwave range and their applications (Saiotek 08 METAMAT, S-PE08UN44)
CALL: GOBIERNO VASCO, SPRI
PARTICIPANTS: UPV/EHU
From: 01/01/2008 *To*: 31/12/2009
PRINCIPAL Researcher (coordinator): Arcady Zhukov

Another Relevant merits

1. Scientific referee of the scientific papers for:
- Journal of Magnetism and Magnetic Materials (>50)
- Sensors and Actuators (A) (3)
- 5th International Workshop on Non-Crystalline Solids (2)
- Intermag 2000 (>10)
- 2nd , 3-d European Conference in Magnetic Sensors &Actuators (EMSA`98, EMSA -2003) (6)

- SMM15 (5)
- J. Appl Phys. and Appl. Phys. lett (regular referee, >20)
- Mat. Sci. Eng.
- International Conference on Magnetism 2003, 2005,
- Joint European Magnetic Symposium. JEMS-01, Grenoble, France August 28-31, 2001(4)

2. Evaluator of ANECA (Spanish agency of quality of Ministry of Education)
3. Participation in the projects with industry:

Project title "New Magnetic, Magnetoelastic and Magnetoresistent Sensors" (*Department of Industry, Agriculture and Fishing of the Basque Government.*

4. Invited Seminars

- University of Nebraska, Center for Materials Research and Analysis (Lincoln, USA) (2001)
- Instituto de Investigaciones en Materiales de la UNAM de Mexico (1999)
- Department of Engineering Physics and Materials Engineering Ecole Polytechnique Montreal (Canada, 2000)
- Center of Advanced European Studies, CAESAR (Germany, Bonn, October 2001)
- MIT(Cambridge, USA)(2001, 2009)
- Moscow State University (Russia, September 2004, April 2009)
- Moscow Technolgical University (Russia, April 2009)
- Kosice University (Slovacia) (Slovak Republic, May 2009)
- Bristol University (UK, September 2009)

5. CO-CHAIRMAN of Joint European Magnetic symposium (San Sebastián, June 2006, 750 participants)

   CO-CHAIRMAN of the symposium "*Recent Research on Novel Magnetic Structures and Their Applications* " (San Sebastián, September 2000)

6. Programme Co-Chairman Of The "International Workshop On Magnetic Wires " (San Sebastián, June 2001)

   Member Of The Organizing Comittees
   - "Magnetism of Nanostructured Phases, MNP Conference" San Sebastián (Septiembre, 1998), responsible for the Publication.
   - Moscow Internacional Symposium on Magnetims "MISM 2005"
   - Internacional conference "EMSA 2006"
   - Eigth Internacional Workshop on Non-Crystalline Solids", Gijon June 20-23, 2006
   - Moscow Internacional Symposium on Magnetism "MISM 2008", June 2008 (Program committee)
   - International Workshop On Magnetic Wires 2008 " (Zumaia, Mayo 2008) (Program committee), 2010 (Bodrum Turkey)
   - "*International conference on* Functional Materials 2007" (Crimea, October 2007) (Program committee)
   - "*International conference on* Functional Materials 2009" (Crimea, October 2009) (Program committee)

- Organizer and session Charimen of "Magnetic Microwave Smart Materials" (Session5A6) of international Conference Progress In Electromagnetics Research Symposium (PIERS) 2009 in Moscow, Agosto 2009

7. Chairing the sessions at international conferences:

-Chairman of the Session "Amorphous and Nanocrystalline Materials" of the 52nd Magnetism and Magnetic Materials Conference
-Chairman of the Session "Soft Magnetic Materials And Emc" Of The International Magnetic Conference(Intermag 2008, Madrid 4-8 May, Intermag 2008)
-Chairman Of The Session "Crystalline, Nanocrystalline, And Amorphous Materials (Ii)" Of The International Magnetic Conference(Intermag 2009, Sacramento Ee.Uu.)
-Chairman Of The Session "Magnetic Materials" Of The The 17-Th Annual International Conference On Composites/Nano Engineering (Icce – 17, Hawaii, July 2009, Usa)
- Co-Chairman Of The Session "As Amorphous And Nanocrystalline Soft Magnets I" Of The 11th Joint MMM–Intermag Conference (Washington, EE.UU., Enero 2010)

8. Member Of The Scientific Tribunal Of The PHD Thesis of Mr. Victor Manuel de la Prida Pidal (Universidad de Oviedo) (Oviedo, May 2000) and Mrs. Leonor Pascual Castaño (Universidad Complutense, Madrid) (Abril 2003), Carmen Miguel (2003, UPV/EHU), Jesus Oliveira (Oviedo 2008)

9. GUEST EDITOR of Physica B. with the Proceedings of Simposium "*Recent Research on Novel Magnetic Structures and Their Applications*" B Physica B 299 (2001).

10. Guest Editor of J.Magn. Magn. Mater . with the Proceedings of the International Simposium "International Workshop On Magnetic Wires " (San Sebastián, June 2001) J.Magn. Magn. Mater 249, 1-2, (2002)

11. Guest Editor of Phys. Status Solidi A with Proceedings of the International Workshop On Magnetic Wires 2008 " (Zumaia, Mayo 2008) Phys. Status Solidi A 206, No. 4, (2009)

12. Guest Editor of Journal of Magnetism and Magnetic Materials 316 (2007) containing proceedings of the Joint European Magnetic symposium (San Sebastián, June 2006, 750 participants)

13. Guest Editor Of Journal Of Nanoscience And Nanotechnology (Special issue devoted to nanomagnetism), Guest Editors O. Chubykalo-Fesenko, J. Gonzalez, B. Hernando, and A. Zhukov, Volume: 8 Issue: 6 Pages: 2729-2729 Published: JUN 2008

14. Member of the Editorial board of the International journal "The Open Physical Chemistry Journal, Open Physical Chemistry Letters and Open Physical Chemistry Reviews" (Editorial - Bentham Science Publishers)

15. Member of the Editorial board of the International journal "Open Materials Science Journal" (Editorial - Bentham Science Publishers)

16. Member of the Editorial board of the journal "Journal of Functional Materials" (Editorial – Science and Technology – in Russian)"

17. Responsible of the post-doctoral stays, of the Visiting Researcher:
   - *Victor Mnauel de La Prida Pidal (Visitng Professor of the Universidad de Oviedo, 3 meses)*
   - *Prof. Anatoly Zvezdin (Profesor Visitante de la Universidad de Moscú, 12 meses), 2008*

   Ikerbasque Fellowship
   - Prof. Larissa Panina (Profesor Visitante de la Universidad de Plymouth, UK , 12 meses), 2009 Ikerbasque Fellowship

18. Member of the Executive Council of the Direction of the Internationalization of the System of Innovation, of Innobasque.

19. Foundator of spin-off Company "TAMAG Iberica S.L." , being main scientific consultant along years all years of existence.

Teaching experience

1. Doctoral courses „Magnetic Materials: Relevant Properties and Applications" (Doctoral Programme "Ciencia y Tecnología de Materiales" of the UPV/EHU 2006-2008)

2. Responsible for the Master course "Nanodevices" (2007-up to now)

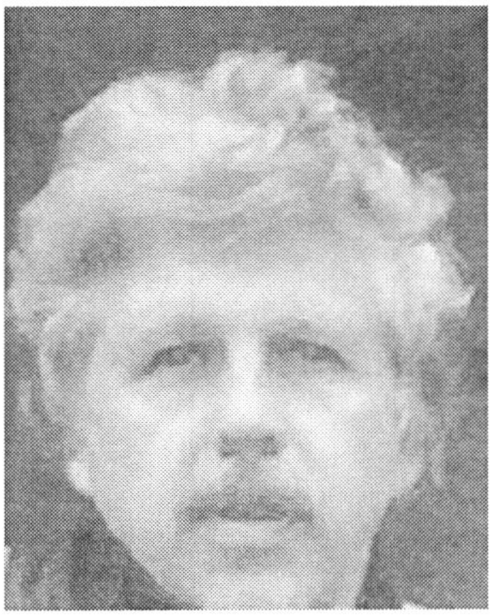

PART 2 - RESEARCH SUMMARIES IN NANOTECHNOLOGY

In: Superconductivity and Condensed Matter … ISBN 978-1-61209-785-5
Editor: Collin J. Eckles

Chapter 33

FLUX PINNING AND AC LOSS STUDIES ON YBCO COATED CONDUCTORS

M. Parans Paranthama[1]n and Venkat Selvamanickam[2]
[1]Oak Ridge National Laboratory, Oak Ridge, TN 37830-8050, US
[2]Superpower, Inc., Schenectady, NY 12304, US

RESEARCH SUMMARY

Major advances have been made in the last 18 years in high-temperature superconductor (HTS) reserach and development, resulting in increased use of HTS materials in commerical and pre-commercial electric-power applications. This new and important book addresses the issues related to flux pinning, AC losses and thick YBCO film growth. Written by top most scientists in the world, it presents the current status and issues related to YBCO coated conductors and the need for further fundamental materials science work in YBCO coated conductor. It will be a useful handbook for years to come.

In: Superconductivity and Condensed Matter ...
Editor: Collin J. Eckles
ISBN 978-1-61209-785-5

Chapter 34

GENERAL THEORY OF SUPERCONDUCTIVITY

Fu-sui Liu and Yumin Hou
Beijing Univ., Beijing, P.R. China

RESEARCH SUMMARY

Based on the latest progress in superconductivity research, the authors form a conventional theory of superconductivity and express BCS paring theory at a higher level. They especially emphasis the importance of the long-range phase coherence for the superconductivity. They also point out some misunderstandings of basic concepts about superconductivity in present textbooks and monographs. They illustrate the correct theory of the high-Tc superconductivity and the application of the high-Tc superconductors to establish a correct and detailed theory regarding this mechanism in different systems.

In: Superconductivity and Condensed Matter …
Editor: Collin J. Eckles
ISBN 978-1-61209-785-5

Chapter 35

IMPROVEMENT OF VORTEX PINNING IN MGB2

W. K. Yeoh, J. Horvat, J. H. Kim and S. X. Dou
Institute for Superconducting and Electronic Materials,
University of Wollongong, Australia

RESEARCH SUMMARY

This book presents current research topics related to the physics of low-dimensional systems such as magnetic nanoparticles-proteins conjugates and transition metal spin clusters (zero-dimensional systems), and both plain and hybrid micro/nano wires embedded in a normal-metal matrix (one-dimensional systems), and both plain and hybrid magnetic/superconducting thin films (two-dimensional systems). The unconventional magnetic and superconducting properties of these systems are experimentally accessed in relation to the preparation conditions and the subsequent chemical/thermal/mechanical treatment. The novel propositions that are discussed in this book could become cornerstones for interdisciplinary applications related to Nephrology and Hemodialysis, Quantum Computing, Electric Power Applications, Magnetic Recording, and Electronics.

In: Superconductivity and Condensed Matter …
Editor: Collin J. Eckles
ISBN 978-1-61209-785-5

Chapter 36

MAGNETISM AND SUPERCONDUCTIVITY IN LOW-DIMENSIONAL SYSTEMS: UTILIZATION IN FUTURE APPLICATIONS

Dimosthenis Stamopoulos
Inst. of Materials Science, Athens, Greece

RESEARCH SUMMARY

The superconductivity of MgB2 has been hidden for nearly 50 years although it has the highest Tc among the intermetallics superconductors. Beside the high Tc, simple crystal structure, large coherence length, high critical field, transparency of grain boundaries to current flow and low normal state resistivity MgB2 is a fascinating topic to study for both large scale application and electronic devices. Moreover, the presence of two-gap superconductivity (ð and ó band) has been theoretically and experimentally established.

However, critical current density of pristine MgB2 drops rapidly in the high magnetic field due to the weak pinning centres and low upper critical field. During the past three years, novel techniques and developments for fabrication of useable MgB2 have been reported, including chemical alloying, irradiation, thermo-mechanical processing techniques and magnetic shielding to improve the critical density, upper critical field and the irreversibility field. Among the studies, atomic substitution, especially using nano-particles, may help in clarification of the superconductivity mechanism thus making it appropriate for practical application. On the other hand, the momentum of enhancing flux pinning using chemical doping is moving to a positive side. Nano-SiC powder is well known for effective improvement of the critical current density.

In this book, a brief introduction of superconductivity in MgB2 is introduced. Further, the basic electronic and magnetic properties of the MgB2 as well as its crystal structure are reviewed. The authors also discuss the preparation method that has been well developed for experimentation on MgB2. Along with that, they review the experimental results of the chemical alloying on MgB2, particularly on the critical current density. In addition, the future prospects of MgB2 and developments for applications in the current superconductivity market are given.

In: Superconductivity and Condensed Matter ... ISBN 978-1-61209-785-5
Editor: Collin J. Eckles

Chapter 37

SUPERCONDUCTING INTERCALATED GRAPHITE

Nicolas Emerya[1], Claire Hérolda and Philippe Lagrange[2]

[1]Laboratoire de Chimie du Solide Minéral – Nancy Université - Université Henri Poincaré, - Vandœuvre-lès-Nancy Cedex – France
[2]Laboratoire de Chimie du Solide Minéral – Nancy Université -Université Henri

RESEARCH SUMMARY

The discovery in 2005 of superconductivity in YbC6 and CaC6, with substantially higher critical temperatures than the previously observed among the family of the graphite intercalation compounds, has largely renewed the interest for these well known lamellar compounds. Indeed, these critical temperatures reach 6.5 and 11.5 K respectively for ytterbium- and calcium-graphite phases. It was consequently interesting to collect all the informations concerning the superconductivity of these compounds from the discovery of this phenomenon observed in the heavy alkali metals graphite intercalation compounds in 1965, insisting particularly on the recent advances in this research field. After a general introduction that describes all the carbon materials, which are extremely various with dimensionalities varying from 3 to 0, leading to their large aptitude for the insertion/intercalation reactions, the authors widely developed the case of graphite: chemical bonds, crystal and electronic structures, anisotropy and ability to become a host structure. The authors insist on its strong anisotropy of chemical reactivity that allows the synthesis of very numerous intercalation compounds. The distinctive features of the intercalation reaction into graphite are reviewed (systematic charge transfer, staging, etc...) and are particularly developed in the case of the donor-type intercalation compounds, among which is precisely observed the superconductivity. For the latter, the various synthesis methods are successively described, showing the best route to use in order to obtain each type of compound. Then the authors review with detail the binary compounds, emphasizing their distinctive crystal and electronic structures and also their transport properties. The authors describ the superconductivity of all the compounds belonging to this family and show this property. In the last part, the authors compare these superconducting binary intercalated graphite compounds with other lamellar superconductor: magnesium diboride. The ternary compounds are then studied, and the poly-layered nature of their intercalated sheets is given special attention. Their distinctive electronic structure is presented and their superconducting properties are described.

In: Superconductivity and Condensed Matter … ISBN 978-1-61209-785-5
Editor: Collin J. Eckles

Chapter 38

UNCONVENTIONAL ELECTROMAGNETICS IN CARBONACEOUS MATERIALS

Sergey G.Lebedev
Institute for Nuclear Research, Moscow, Russia

RESEARCH SUMMARY

This book presents some new results in the studies of anomalous electromagnetic properties of thin granular carbon films.

The book describes the study of some new anomalous electromagnetic effects in graphite-like thin carbon films. These are:

- The fast switching (10-9 sec) of electrical conductivity
- The detection of microwave radiation and its temperature dependence
- The oscillations of film stack magnetization in a magnetic field of 1-5 T
- The generation of optical radiation during the spasmodic switching of conductivity

In: Superconductivity and Condensed Matter ... ISBN 978-1-61209-785-5
Editor: Collin J. Eckles

Chapter 39

THE HIERARCHIC THEORY OF LIQUIDS AND SOLIDS: COMPUTERIZED APPLICATIONS FOR ICE, WATER AND BIOSYSTEMS

Alex Kaivarainen
Univ. of Turku, Dept. of Physics, Turku, Finland

RESEARCH SUMMARY

This book presents a review of an original Hierarchic theory of condensed matter, general for liquids and solids and its numerous applications. Computer programs based on a new theory were used for comprehensive simulations of water and ice physical properties and validation of the theory. Condensed matter is considered as a system of 3D standing waves (collective excitations) of different nature: thermal de Broglie waves, IR photons and thermal phonons. Quantitative interrelation between microscopic, mesoscopic (as intermediate) and macroscopic properties of condensed matter were found out.

New theories of total internal energy, including contributions of kinetic and potential energies, heat capacity, surface tension, vapor pressure, thermal conductivity, viscosity and self-diffusion are described. Hierarchic theory of osmotic pressure, based on new state equation, new theories of light refraction, Brillouin light scattering and Messbauer effect are presented also in article and compared with available experimental data for water and ice. Lot of hidden parameters, inaccessible for experiment, describing the dynamic and spatial properties of 24 quantum collective excitations of matter, can be calculated also, as demonstrated on examples of water and ice. Total number of physical parameters of liquids and solids in wide T-interval, including that of phase transitions, to be possible to evaluate using CAMP computer program, is about 300. The agreement between theoretical and available experimental results is good. The evidence of high-T mesoscopic Bose condensation (mBC) in water and ice in form of coherent clusters is obtained. The new mechanisms of the 1st and 2nd order phase transitions, related to such clusters formation/melting, their assembly/disassembly and symmetry change is proposed. The influence of weak magnetic field on water is demonstrated experimentally and explained in the framework of new theory.

Theory unifies dynamics and thermodynamics on microscopics, mesoscopic and macroscopic scales in terms of quantum physics. The idea of new optoacoustic device: Comprehensive Analyzer of Matter Properties (CAMP) with huge informational possibilities, its possible configurations and applications are described.

In: Superconductivity and Condensed Matter ...
Editor: Collin J. Eckles

ISBN 978-1-61209-785-5

Chapter 40

PHASE TRANSITIONS INDUCED BY SHORT LASER PULSES

Georgy A. Shafeev
Wave Research Center, A.M. Prokhorov General Physics Institute,
Moscow, Russian Federation

RESEARCH SUMMARY

The book presents the recent advances in studies of phase transitions induced by short laser pulses in solids and the interface solid-liquid. The review is presented of previous and new results of experimental and theoretical description of phase boiling phenomenon under the action of short laser pulses. Time-resolved characterization of the phase transitions induced by short laser pulses in metals in confined geometry is described in details, and the results are compared with classical equation of states of the materials in sub- and super-critical range of parameters. Phase transitions in the films of cryogenic molecular compounds doped with different impurities are thoroughly studied with respect to their use in MALDI and MAPLE processes. Modeling and new experimental data are described on explosive boiling of liquid under laser exposure of its free surface. Novel experimental results are presented on instabilities at the solid-liquid interface exposed to pico- and femtosecond laser pulses. Morphological and optical properties of resulting nanostructures periodic relief are described. Theoretical modeling of instabilities and structures formation under laser exposure is given on the basis of defect-deformational instability. Modeling results are found to be in good agreement with experimental data. The book is of interest for scientists, post-graduate students and engineers working in the field of laser-matter interaction.

In: Superconductivity and Condensed Matter ... ISBN 978-1-61209-785-5
Editor: Collin J. Eckles

Chapter 41

EXCHANGE BIASED AND PLAINSUPERCONDUCTING-FERROMAGNETIC LAYEREDHYBRIDS AND THEIR POSSIBLE APPLICATIONS

D. Stamopoulos*[†]*, E. Manios and M. Pissas
Institute of Materials Science, NCSR "Demokritos", 153-10, Aghia Paraskevi, Athens, Greece

RESEARCH SUMMARY

It is generally believed that superconductivity (SC) and ferromagnetism (FM) are two antagonistic long-range phenomena since the exchange field is detrimental to conventional singlet-pairing mechanisms. Here, we show that under specific circumstances these two phenomena may become cooperative so that their synergy may promote the transport properties of SC-FM layered hybrids. We have studied systematically the magnetic and transport behavior of such *exchange biased* and *plain* layered hybrids that are comprised of $Ni_{80}Fe_{20}$and low-T_cNb as the FM and SC ingredients. Four structures have been studied: exchange biased FM-SC-FM trilayers (TLs), exchange biased FM-SC bilayers (BLs), and also their respective plain ones. Detailed magnetization data on the *longitudinal* and *transverse* magnetic components are presented for both the *normal* and *superconducting* states. These data are compared to systematic transport measurements that were performed as a function of both temperature and magnetic field. For both the exchange biased and plain TLs the transport data reveal a major magnetoresistance effect that emerges only in the superconducting state of the Nb interlayer. In contrast, in the BLs the magnetoresistance effect is only minor whether they are exchange biased or not. Also, the dynamic behavior of the hybrids is presented through complete I-V characteristics. The direct comparison of the exchange biased TLs and BLs with the plain ones enable us to reveal all the underlying parameters that influence their transport properties. Primarily, the underlying mechanism motivating the extreme magnetoresistance peaks in the TLs relates to the suppression of superconductivity mainly due to the stray-field induced transverse magnetic coupling of the

[†] E-mail address: densta@ims.demokritos.gr

two FM layers as the out-of-plane rotation of their magnetizations takes place across the coercive field where the emerging transverse stray fields penetrate completely the Nb interlayer. Atomic force microscopy is employed in order to examine the possibility that such magnetostatic coupling could be promoted by interface roughness. Secondarily, the relative in-plane magnetization configuration of the outer FM layers also exerts a minor contribution on the SC interlayer so that a spin-dependent scattering mechanism could also be invoked as in the giant magnetoresistance effect that is observed in conventional TLs of normal metal interlayer. Most importantly, since the exchange bias may directly control both the in-plane and out-of-plane magnetic order the magnetoresistance peaks may be tuned at will. Also, we reveal that clear fingerprints of exchange bias, such as the training effect, are well preserved in the superconducting state. Our experiments not only point out the need for a new theoretical treatment of FM-SC hybrids but also direct us toward the design of efficient supercurrent-switch elemental devices.

In: Superconductivity and Condensed Matter ...
Editor: Collin J. Eckles

ISBN 978-1-61209-785-5

Chapter 42

TRANSPORT PHENOMENA DUE TO STRUCTURAL INHOMOGENEITY IN SUPERCONDUCTING CUPRATES DOPED WITH CATIONS

Katsukuni Yoshida
Graduate School of Energy Science, Kyoto University, Kyoto 606-8501, Japan

RESEARCH SUMMARY

In this review article, we describe the electrical and diamagnetic properties which are characteristic of such nanostructural inhomogeneity as introduced by dopant like Pr into high-T_c cuprates. The main cuprates we focus here are of the type $R\mathrm{Ba_2Cu_3O_7}$ (shortly R-123) with rare earth element R, which belongs to the class of the critical temperatures (or onset) T_c=90~96 K, as represented by $\mathrm{YBa_2Cu_3O_7}$ (Y-123). Until recently, little attention has been paid to such structural inhomogeneity in describing a variety of transport phenomena in the doped cuprates like $R_{1-x}\mathrm{Pr}_x\mathrm{Ba_2Cu_3O_{7-\delta}}$ (RPr-123), and most of them seem to have been approached in terms of the variation in the overall concentration of mobile holes or breaking of hole-pairs. Here we describe the phenomena in terms of the nanostructural inhomogeneity. In particular, a main issue on the inhomogeneous structure in the doped cuprates will be focused on the percolation processes, where crystal unit cells of R-123 and Pr-123 constitute a mixed crystalline solid solution. The other effects on different dopant like Nd or Sr when occupying Ba-site will also be described.

In this article the following subjects will be described in reference to the structural behavior in the doped cuprates.

1. Polycrystalline composites of two phases Y-123 and Pr-123 and the effect of interfusion between them on the transport properties (Sec.2).
2. Non-ohmic transport phenomena due to the percolative inhomogeneity in Pr-doped cuprates $R_{1-x}\mathrm{Pr}_x\mathrm{Ba_2Cu_3O_{7-\delta}}$ (RPr-123) (Sec. 3).

3. Nonclassical percolation process and the superconductor-insulator transition in the mixed crystalline solution Pr-doped Y-123 ($Y_{1-x}Pr_xBa_2Cu_3O_{7-\delta}$) (Sec. 4).
4. Effects of cation substitution at Ba site in doped- $RBa_2Cu_3O_{7-\delta}$ (Sec.5).

In: Superconductivity and Condensed Matter ...
Editor: Collin J. Eckles
ISBN 978-1-61209-785-5

Chapter 43

EFFECTS OF PARTICLE IRRADIATION ON METALLIC MgB_2 SUPERCONDUCTOR

A. Talapatra, S. K. Bandyopadhyay, Pintu Sen and P. Barat
Variable Energy Cyclotron Centre, 1/AF, Bidhan Nagar, Kolkata 7000064

RESEARCH SUMMARY

MgB_2 is a two-band superconductor with highest T_c among all the metallic superconductors. It has also high J_c. These properties make it a potential candidate for its application in technological purpose like high field superconducting magnets for high energy research, cryogenerator based magnets, detectors etc. For this purpose, it is important to see the effects of particle irradiation on MgB_2. Moreover, the irradiation study of MgB_2 gives the idea about the effects of disorder in multi-band superconductor. In this article we present the effects of ions and neutron irradiation on normal and superconducting properties of MgB_2 and compared it with conventional and high temperature superconductors. In section one, we give a preliminary introduction to MgB_2 and two-band superconductor. Section two deals with general introduction to radiation damage in solids and its effects in superconductors. In section three we discuss about the resistivity and T_c of irradiated and pristine MgB_2 and compare with conventional and high temperature superconductors. Section four deals the irradiation effects on H_{c2} due to irradiation. Critical current and irreversibility field is discussed in section five.

In: Superconductivity and Condensed Matter ...
Editor: Collin J. Eckles
ISBN 978-1-61209-785-5

Chapter 44

THE DEVELOPMENT OF NANOPARTICLES DOPING IN MELT-PROCESSED GD123 SUPERCONDUCTORS

***Caixuan Xu*[1,2], *Anming Hu*[3], *Masaki Ichihara*[4], *Naomichi Sakai*[2], *Izumi Hirabayashi*[2] and *Mitsuru Izumi*[1]**

[1]Department of Marine Electronics and Mechanical Engineering, Tokyo University of Marine Science and Technology, Tokyo 135-8533, Japan

[2] Superconductivity Research Laboratory, International Superconductivity Technology Center, Tokyo 135-0062, Japan

[3]Department of Physics, University of Waterloo, Waterloo, ON, N2L 3G1, Canada

[4]Institute for Solid State Physics, University of Tokyo, Chiba 277-8581, Japan

RESEARCH SUMMARY

Melt-processed $GdBa_2Cu_3O_{7-\delta}$ (Gd123) superconductors were fabricated in air by doping nanoparticles which with an average size of 50 nm. The doping effect on microstructure and superconducting properties was investigated by scanning electron microscopy (SEM) / transmission electron microscopy (TEM)/ atomic force microscopy (AFM) and SQUID. SEM results show that the distribution of Gd_2BaCuO_5 (Gd211) particles is inhomogeneous around the seeding site of single domain, induced by the addition of nano-particles. Particles with size less than 50 nm were observed by TEM inside Gd123 single domain, while the typical size of fined Gd211 particles is about 100 nm. In addition, the nanostripes have been observed by the AFM in the samples with doping nanosized ZnO particles. Critical current density (J_c) was enhanced by the doping proper amount nanoparticles, especially for ZrO_2 and SnO_2. The highest J_c of 100,000 A/cm^2 was presently achieved in the bulk with 0.4 mol % ZrO_2, under self-field at 77 K. Excess amount addition rather suppresses superconducting properties, for instance, 10 mol % ZnO doping will completely destroyed the superconductivity. The optimum addition amount is 0.2-0.8 mol % for ZrO_2 and 0.4-0.8 mol % for SnO_2, ratio to Gd123, respectively.

In: Superconductivity and Condensed Matter …
Editor: Collin J. Eckles
ISBN 978-1-61209-785-5

Chapter 45

COHERENT OSCILLATIONS IN SUPERCONDUCTING COLD FERMI ATOMS AND THEIR APPLICATIONS

Razvan Teodorescu
T-13/CNLS, Los Alamos National Laboratory, Los Alamos, NM 87545, US

RESEARCH SUMMARY

Recent achievements in experiments with cold fermionic atoms indicate the poten-tial for developing novel superconducting devices which may be operated in a wide range of regimes, at a level of precision previously not available. Unlike traditional, solid-state superconducting devices, the cold-atom systems allow the fast switching on of the BCS phase, and the observation of non-equilibrium, coherent oscillations of the order parameter. The integrable and non-linear nature of the equations of motions makes this operating regime particularly rich in potential applications, such as quan-tum modulation and encoding, or nonlinear mixing of quantum coherent oscillations, to name only two.

From a mathematical point of view, such systems can be described using the Knizhnik-Zamolodchikov-Bernard equation, or more generally, by the matrix Kadomtsev-Petviashvilii integrable hierarchy. This identification is particularly use-ful, since it allows a direct application of the known non-linear phenomena described by particular solutions of these equations. Other important features of this formula-tion, such as the relation to the quantum Calogero model, also have relevant physical interpretations. In this work, a complete description of these relationships is presented, along with their practical consequences.

In: Superconductivity and Condensed Matter ...
Editor: Collin J. Eckles
ISBN 978-1-61209-785-5

Chapter 46

A New Approach to the Electron State in Transition Metal Oxides: Covalent σ and π Bonding between Ions and Two Types of Electrons

M. V. Krasinkova*
Ioffe Physical-Technical Institute of Russian Academy of Sciences
Polytekhnicheskaya 26, 194021 St. Petersburg, Russia

Research Summary

It has been shown that many unusual properties of transition metal oxides, including such collective phenomena as magnetic ordering in manganites and high-temperature superconductivity in cuprates, can be explained if a mixed ionic-covalent nature of the transition metal-oxygen bond and polarization of anions by cations are taken into consideration. It has been found that the role of polarization in these materials is not restricted to formation of electric dipole moments, the polarization results in the appearance of a mixed bond covalent component including covalent σ and π type interionic bonds to which electrons of two types (σ and π) correspond. The σ electrons participate in formation of localized one-electron interionic bonds and form a localized spin-ordered electron lattice which is responsible for dielectric properties of these materials and their magnetic ordering. The dielectric state can be destroyed by doping giving rise to π bonding between ions, additional to σ bonding. The π electrons form Wigner lattices, which creates conditions for charge transfer (conductivity and superconductivity) in these materials.

* E-mail address: marina.shuv@mail.ioffe.ru

In: Superconductivity and Condensed Matter ...
Editor: Collin J. Eckles

ISBN 978-1-61209-785-5

Chapter 47

VORTEX THEORY OF INHOMOGENEOUS SUPERCONDUCTORS

B. J. Yuan
Department of Physics, Yuxi Normal University, Yuxi, Yunnan Province, P. R. China

RESEARCH SUMMARY

Combining a number of well-known theories of conventional superconductors [1–6], which include Abrikosov vortex theory, Gorkov formalism, Extended Ginzberg-Landau theory, and de Gennes-Werthamer proximity coupling theory, a general vortex theory for inhomogeneous superconductors is proposed [7–9]. This general vortex theory is applied to three characteristic and simple geometries, a bulk superconductor with finite size, a thin film in parallel applied filed and a special two-components superlattice, respectively. In each of the three cases, spatial form of order parameter and magnetic field in the geometry are derived, which lead to an analytical expression of Gibbs free energy. For a finite bulk superconductor, the eigenvalue spectrum in central region of the bulk geometry is in similar to the case of an infinite geometry with lower symmetry. A vortex lattice in the central region prefers a hexagonal structure if only minimum free energy is concerned, however, when effects of repulsion between flux lines or arrayed defects are included, symmetry of a vortex lattice varies. For a thin film geometry in a parallel field, the solution shows that the inhomogeneity caused by limited thickness of a film enhances extended GL parameter, κ_2, which proves that thin film superconductor tends to be type II superconductor. For a two-components (NS) superlattice in a parallel applied filed, we find that, in addition to a slowly spatial varying pair amplitude nucleating in N layers, a highly condensed pair amplitude confined in the other component layers (S) can also simultaneously satisfy both a minimum eigenenergy requirement and a minimum free energy requirement. The formation of a vortex lattice either in N or S layers is determined by the competition of the two mechanisms. The analysis to different geometries shows that there are two energy criteria in forming a vortex lattice. Both energy criteria are affected explicitly or implicitly by the inhomogeneities of a superconductor.

In: Superconductivity and Condensed Matter ...
Editor: Collin J. Eckles

ISBN 978-1-61209-785-5

Chapter 48

SYNTHESIS, STRUCTURE, SUPERCONDUCTING AND MAGNETIC PROPERTIES OF $RuSr_2LnCu_2O_{8-\delta}$ AND $RuSr_2(Ln,Ce)_2Cu_2O_{10-\delta}$: A STRUCTURAL ASPECTS

*G. M. Kuz'micheva**
Department of Solid State Physics and Chemistry,
M. V. Lomonosov State Academy of Fine Chemical Technology,
86 Vernadsky prosp., Moscow 117571, Russia

RESEARCH SUMMARY

All known high-T_c superconducting compounds (HTSC) are the derivatives of $CaTiO_3$ (cubic perovskite) structural type. An ever growing variety of HTSC with T_c from ~30 to 134K has been synthesized since the discovery of superconductivity in the cuprate system $La_{1-x}Ba_xCuO_4$. Over the past few years, researchers have focused more on rutheno-cuprate compounds, in which the coexistence of superconductivity and magnetic order has been found. Two cations, Ru and Cu, with variable formal charges and unlike properties, are highlighted in these systems, as both fulfill different functions. This paper reviews the relation between the synthesis conditions, the composition of prepared phases, some characteristics of magnetic and superconducting properties and structure parameters of rutheno-cuprates.

* Corresponding author. E-mail address: galkuz@orc.ru; Tel:+7-095-2480762; Fax:+7-095-4348711

In: Superconductivity and Condensed Matter ... ISBN 978-1-61209-785-5
Editor: Collin J. Eckles

Chapter 49

HIGH T_C SUPERCONDUCTING FIBERS PROCESSED BY CONVENTIONAL AND ELECTRICAL ASSISTED LASER FLOATING ZONE

***F. M. Costa*[1], *M. F. Carrasco*[1], *R. F. Silva*[2] and *J. M. Vieira*[2]**

[1]Department of Physics, University of Aveiro, 3810-193 Aveiro, Portugal
[2]Department of Ceramics and Glass Engineering, CICECO, University of Aveiro, 3810-193 Aveiro, Portugal

RESEARCH SUMMARY

In the last years there has been an ongoing interest in the development of new processes to produce crystallographic alignment in bulk materials. The high axial temperature gradients and quenching rates of the laser floating zone (LFZ) method make it a powerful technique to grow highly textured materials. In polycrystalline oxide superconductors, one of the major drawbacks that affect the critical current density is the presence of high-angle grain boundary weak links. The LFZ technique permits the epitaxial growth of superconducting phases on the surface of the primary cuprate phase, during solidification, leading to a favorable orientation for current transport along the fiber axis direction.

In this work, the LFZ technique is applied to grow superconducting fibers of the Bi-Sr-Ca-Cu-O system. Being a non-equilibrium process, metastable dendritic phases often appear in the as-grown fiber and intergrowth domains develop within the superconducting matrix crystals. The analysis of the solidification interface from the quenched melted zone towards the solid fiber allows the establishing of the crystallization course. The modification of the effective distribution coefficients of the different cations puts in evidence the strong effect of the pulling rate on solute trapping. After annealing, the metastable phases disappear leading to the equilibrium structure consisting of high number CuO_2 layered superconducting phases.

A new method for controlling the solidification of superconducting materials is proposed by applying an electric current through the solid/liquid interface during the LFZ process. The electrical assisted laser floating zone (EALFZ) technique improves the texture degree and the quality of the intergranular contacts. In the presence of an electric current, the solute transport outcomes from the balance between the local solute gradients resulting from constitutional supercooling and electromigration effects. Phase development and effective distribution coefficients are strongly modified in comparison to the conventional LFZ process. This new EALFZ process makes possible the tailoring of the fiber microstructure, namely the nature and shape of crystalline phases developed, regarding an improvement of the superconducting properties, mainly the critical current density.

In: Superconductivity and Condensed Matter ...
Editor: Collin J. Eckles
ISBN 978-1-61209-785-5

Chapter 50

SPIN GLASS BEHAVIOR ON RUTHENO CUPRATES

C.A. Cardoso and F.M. Araujo-Moreira
Grupo de Materiais e Dispositivos, CMDMC, Departamento de F´ısica, UFSCar, Caixa Postal 676, 13565-905 , S˜ao Carlos-SP, Brazil
V.P.S. Awana and H. Kishan
National Physical Laboratory, K. S. Krishnan Marg.,New Delhi 110012, India
E. Takayama-Muromachi
Superconducting Materials Center, National Institute for MaterialsScience, 1-1 Namiki, Tsukuba, Ibaraki 305-0044, Japan
O.F. de Lima
Instituto de F´ısica ”Gleb Wataghin”, UNICAMP, 13083-970,Campinas-SP, Brazil

RESEARCH SUMMARY

Strong evidences of a spin glass phase on polycrystalline $RuSr_2Ln_{1.5}Ce_{0.5}Cu_2O_{10-\delta}$ (LnRu −1222, Ln = Y, Ho and Dy) is pro-vided by ac susceptibility (χ_{ac}) and dc magnetization measurements. The fingerprint of a spin glass transition, the frequency-dependent cusp in χ_{ac} vs. T measurements at low magnetic field, is observed for all LnRu−1222samples. The change in the cusp position with frequency follows the Vogel-Fulcher law, which is commonly accepted to describe a spin glass with magnetically interacting clusters. The strong supression of the cusp in the ac susceptibility measurement in the presence of magnetic fields as low as 500 Oe, and the results of thermoremanent magnetization (TRM) and isothermal remanent magnetization (IRM) measurements are also evidences of a spin glass behavior. By comparing these results with measurements on a YRu −1212 sample we conclude that the glassy behavior is characteristic of the Ru−1222phase, not being observed for the studied YRu −1212 sample. We suggest that the spin glass transition occur only in the Ru−1222 phase possibly due to the larger number of oxygen vacancies present in these compounds.

In: Superconductivity and Condensed Matter ...
Editor: Collin J. Eckles
ISBN 978-1-61209-785-5

Chapter 51

ROLE OF THE VORTEX SOLID TOPOLOGY IN A FIRST-ORDER LIQUID-SOLID PHASE TRANSITION

M. Menghini[1], Yanina Fasano[2*], F. de la Cruz[2], S.S. Banerjee[3], Y. Myasoedov[3], E. Zeldov[3], C.J. van der Beek[4], M. Konczykowski[4] and T. Tamegai[5]

[1]Division of Physics and Astronomy, Faculty of Sciences, Vrije Universiteit, De Boelelaan 1081, 1081HV Amsterdam, The Netherlands

[2]Instituto Balseiro and Centro At´omico Bariloche, CNEA, Bariloche, 8400, Argentina

[3]Department of Condensed Matter Physics,Weizmann Institute of Science, Rehovot 76100, Israel

[4]Laboratoire des Solides Irradi´es, CNRS UMR7642 and CEA-CSM-DRECAM,Ecole Polytechnique, F-91128 Palaiseau Cedex, France

[5]Department of Applied Physics, The University of Tokyo, Hongo,Bunkyo-ku, Tokyo 113-8656,and CREST Japan Science and Technology Corporation (JST), Japan

RESEARCH SUMMARY

Using a combination of magnetic decoration and differential magneto-optical tech-niques we have studied the interrelation between the order of the vortex liquid-solid phase transition and the corresponding structure of the vortex solid phase in $Bi_2Sr_2CaCu_2O_8$crystals with different doses of columnar defects (CDs). The mag-netic decoration makes evident that when the number of vortices exceeds that of CDs, the vortex solid state is a heterogeneous system containing two species of vortices: A network of grain boundaries associated with vortices pinned on CDs surrounding regions free of columnar defects where Abrikosov microcrystals solidify. On the other hand, when the number of vortices is smaller than that of columnar defects the struc-ture of the vortex solid is amorphous. The vortex liquid-solid phase transition as deter-mined by differential magneto-optical experiments shows that the first-order melting phase transition does not require a structural symmetry break as usually found in na-ture. Comparative analysis of the topology of the vortex solid and the characteristic of the corresponding melting line reveals a new correlation between the melting mecha-nisms and the percolation of superconductivity in the heterogeneous vortex solid.

In: Superconductivity and Condensed Matter ...
Editor: Collin J. Eckles
ISBN 978-1-61209-785-5

Chapter 52

GROWTH KINETICS OF Yb:$YAl_3(BO_3)_4$ AND $NdAl_3(BO_3)_4$ EPITAXIAL LAYERS

E. A. Volkova and N. I. Leonyuk
Department of Crystallography and Crystal Chemistry, Geological Faculty, Moscow State University, Moscow 119992/GSP-2, Russia
A. V. Mokhov
Institute of Ore Deposits Geology, Petrography, Mineralogy and Geochemistry, Moscow 109017, Russia Japan

RESEARCH SUMMARY

Yb:$YAl_3(BO_3)_4$ (Yb:YAB) and $NdAl_3(BO_3)_4$ (NAB) thin films were grown from $K_2Mo_3O_{10}$-B_2O_3 based fluxed melts using substrates made of pure $YAl_3(BO_3)_4$ crystals. Homogeneity, micromorphology and composition of Yb:YAB and NAB thin films were studied in dependence on supersaturation and temperature regime of fluxed melts. Relations between the growth rate V and the relative supersaturation β were determined and compared with previous measurements for YAl- and YFe-borate single crystals. It was shown that experimental curves V vs β of NAB and Yb:YAB (Yb-5 at%) can be described by parabolic law, while linear relation was revealed for 10 at% concentration of Yb in YAB epitaxial layers. It was found that primarily volume and surface diffusion processes occur simultaneously, although kinetic limitation of the growth rate dominates on final stages of crystallization. The characteristic morphological features of Yb doped YAB single crystal layers result in growth spirals and vicinal hillocks. Besides, cusps and corrugations are observed on main faces. Yb concentration in grown films slightly increases from 0.09 to 0.11 with an increasing supercooling/supersaturation, but these variations are not dramatic at this stage. The surface structure of NAB layers strongly depends on the growth temperature. These layers were deposited by layer-by-layer (Frank – van der Merve) mode. Nevertheless, the transition from three- to two-dimensional mechanism is observed at high supersaturations.

In: Superconductivity and Condensed Matter ...
Editor: Collin J. Eckles

ISBN 978-1-61209-785-5

Chapter 53

FLUCTUATION AND GAUGE EFFECTS ON THE CRITICAL BEHAVIOR OF SUPERCONDUCTORS

Diana V. Shopova* and Dimo I. Uzunov
CP Laboratory, Institute of Solid State Physics,Bulgarian Academy of Sciences, BG-1784 Sofia, Bulgaria, andMax-Plank-Institut f¨ur Physik komplexer Systeme,N¨othnitzer Str. 38, 01187 Dresden, Germany

RESEARCH SUMMARY

Gauge effects on the fluctuation properties of the normal-to-superconducting phase transition in bulk and thin film superconductors are reviewed. Similar problems in the description of other natural systems (liquid crystals, quantum field theory, early uni-verse) are also discussed. The relatively strong gauge effects on the fluctuations of the ordering field at low spatial dimensionality Dand, in particular, in thin (quasi-2D) films are considered in details. A special attention is paid to the fluctuations of the gauge field. It is shown that the mechanism, in which these gauge fluctuations affect the phase transition order and other phase transition properties varies with the variation of spatial dimensionality D. The problem for the experimental confirmation of theo-retical predictions about the order of phase transitions in gauge systems is discussed. Related topics: gauge effects on the critical behavior of unconventional superconduc-tors, disorder, quantum fluctuations in a close vicinity of ultra-low phase transition temperatures , are also briefly discussed.

* E-mail address: sho@issp.bas.bg. (Corresponding autor)

In: Superconductivity and Condensed Matter ...
Editor: Collin J. Eckles
ISBN 978-1-61209-785-5

Chapter 54

EEM: THE EXCITON ENHANCEMENT MECHANISM THEORY AND EXPERIMENTAL EVIDENCE OF OPTICALLY ENHANCED TC IN HIGH TC SUPERCONDUCTORS

K. W. Wong **and** ***S. Curatolo***[†]
CZT Inc., Lawrence, KS 66049, US

RESEARCH SUMMARY

The Excitonic Enhancement Mechanism is proposed as the theory for superconductivity in some novel hole carrier systems, such as M_3C_{60}, cuprate high temperature superconductors as well as the most recently discovered Boron doped Silicon. It is suggested that because excitons are like hydrogen atoms and possess excited energy levels, these Cooper pair exchange particles can be excited by suitable infrared light, resulting in large enhancement of the superconductivity transition temperature. Experimental data on the EEM Tc enhancement was shown for bulk and thin film YBCO. With the optimum YBCO bulk system (inside an open liquid Nitrogen cryogenic chamber), we show the enhancement by observing the diamagnetism before and after IR light was applied. The transition temperature was shown to increase ~80°K, from $93°K\ to\ \geq\ 175°K$ exactly as predicted by the EEM theory. The results of these original measurements were verified on an oxygen deficient YBCO thin film by measuring the transition temperature using AC voltage versus temperature measurement (inside a closed Helium cryogenic chamber) carefully obtained by using a 100Khz DSP lock-in-amplifier, before and after IR light was applied. The transition temperature was shown to increase by ~17°K, from 69.5°K to 86.5°K. The increase in Tc, although small, is shown to be exactly as the value calculated by the same EEM theory. It is therefore without a doubt that these results confirm EEM as the theory for cuprates and its application to high Tc superconductors.

[*] Kai Wai Wong, Professor Emeritus University of Kansas
[†] Corresponding author: Susana Curatolo, President CZT Inc. czt@sunflower.com

In: Superconductivity and Condensed Matter ...
Editor: Collin J. Eckles
ISBN 978-1-61209-785-5

Chapter 55

SPONTANEOUSLY BROKEN SYMMETRY AND INHOMOGENEOUS SUPERCONDUCTING STATES IN FERROMAGNET/SUPERCONDUCTOR NANOSTRUCTURES

Mansur G. Khusainov[1-3,*]***, Marat M. Khusainov***[4]
and Yurii N. Proshin[1,3,†]
[1]Kazan State University, Kazan 420008, Russia
[2]Kazan State Tupolev Technical University, Chistopol' 422981, Russia
[3]Max-Planck-Institute for the Physics of Complex Systems, Dresden 01871, Germany
[4]Zavoisky Physical-Technical Institute of RAS, Kazan 420029, Russia

RESEARCH SUMMARY

We derive a new boundary-value problem for the ferromagnetic metal/superconductor (FM/S) nanostructures, where superconductivity is a superposition of the BCS pairing with zero total momentum in the S layers and the Fulde-Ferrell-Larkin-Ovchinnikov (FFLO) pairing with nonzero 3D coherent momentum **k** in the FM layers. It is shown that processes of mutual transformation between the BCS and FFLO pairs at the S/FM boundary go via the unrealized *virtual states* (FFLO and BCS, correspondingly). These processes occur as the *Umklapp processes* during which excess coherent pair momentum **k** becomes confined in the FM layer. This fact is reflected in the significantly new boundary conditions for the Eilenberger equations. On this base we originally investigate the interplay between the BCS and 3D FFLO states in the pure FM/S bilayers. In the FM/S/FM trilayers two novel π *phase superconducting states with electron-electron repulsion* in the FM layers are predicted. The 3D modulated FFLO states are possible in such trilayers only in presence of external magnetic field at suitable parameters of the FM and S layers. In the FM/S superlattices there

[*] E-mail address:mgkh.kgtu@mail.ru
[†] E-mail address: Yurii.Proshin@ksu.ru

are also two π phase magnetic states (0π and $\pi\pi$) with compensation of the exchange field paramagnetic effect. This fact allows us to explain a surprisingly high $T_c \sim 5$K in the short period Gd/La superlattice and to predict the sign and value of the electron-electron interaction in the Gd metal.

In: Superconductivity and Condensed Matter ...
Editor: Collin J. Eckles
ISBN 978-1-61209-785-5

Chapter 56

DDW ORDER AND ITS ROLE IN CUPRATES

Angsula Ghosh

RESEARCH SUMMARY

The high-T_c cuprates have been a focus of intense study since its discov ry in 1986. The 20th anniversary of the discovery of this material is marked by a partial understanding of its properties. They are far from being fully understood. The cuprates not only demonstrate a very high T_c, but, also exhibit various anamolies in their super conducting and the normal state properties. The pairing symmetry of the Cooper pairs was a topic of intense debate for more than a decade. A new class of phase sensitive tests as well as various other techniques for energy-gap measurements have revealed a predominantly d-wave pairing symmetry in the cuprates unlike the s-wave states in conventional superconductors. At present one of the most peculiar and controversial topics in these materials is the phenomenon of pseudo gap in the normal state of the cuprates. A depletion of the low-energy density of states is observed in the under doped regime of these materials by various experimental measurements like the optical conductivity, ARPES, c-axis tunneling, specific heat etc. The pseudo gap has also been observed to have the same k-dependence as that of the super conducting gap. A number of theoretical scenarios have been proposed to explain the origin of this phase. It was initially believed to be a crossover phenomenon where the pseudo gap had been ascribed as a natural consequence of local, dynamic pairing correlations in the normal state of low-density small pair size superconductors. A part from the other suggestions like the phase fluctuations of the superconducting order parameter in the pseudo-gap regime, recently, the d-densiywave (DDW) scenario has been proposed as a strong candidate to explain the pseudogap. In fact, various theoretical works on DDW have been successful in explaining some of the salient features of this phase. A number of experiments are also consistent with this proposal. In this article we will present a brief overview of the phase diagram of the cuprates and the present experimental situation followed by a discussion on the present status of the DDW order as a candidate for the pseudogap. We will present a comparison between the DDW order and the other scenarios. We will also explain, in short, the conventional density-wave order and its difference with the unconventional or DDW order. We will discuss the theoretical findings of the DDW order in the limelight of the available experimental predictions.Some proposals of experimental measurements which might play an important role to strengthen the candidature of DDW order will also be considered.

In: Superconductivity and Condensed Matter ... ISBN 978-1-61209-785-5
Editor: Collin J. Eckles

Chapter 57

MODEL OF SUPERCONDUCTIVITY IN THE SINGULAR FERMI LIQUID

Ryszard Gonczarek and Mateusz Krzyzosiak*
Institute of Physics, Wrocław University of Technology,Wybrze˙nskiego 27,
50-370 Wrocław, Polandze Wyspia´

RESEARCH SUMMARY

Superconductivity in a system of strongly correlated electrons treated as a singular Fermi liquid (SFL) is studied. In the system the scattering amplitude for particles with the same momenta is divergent. The qualitatively new ground state and other eigen-states for the superconducting phase are found. Thermodynamic functions are derived within the Bogolubov method. Analytical formulae for the free energy as well as the gap equation and the equation for the chemical potential are presented. The gap equ-ation and the equation for the chemical potential are found to be mutually correlated and have to be simultaneously solved in a self-consistent manner. The numerical solu-tion for the case of the *s*-wave paired SFL is discussed. It is found that the energy gap and the chemical potential are doubly-valued in the subcritical region, however only the upper branches are thermodynamically stable.

According to the common property of temperature-doping phase diagrams for high-$T_c$superconductors in the underdoped regime, two characteristic temperatures can be identified in the SFL model. The higher one is the temperature T_* at which phase-incoherent Cooper pairs appear and the pseudogap emerges. The other one is T_c, lower than T_*, at which the Cooper pairs become phase-coherent and form a su-perconducting condensate. The obtained results evidence that the SFL system displays some features of the superconducting state also in the supercritical temperature range, $T_c<T<T_*$: The non-zero value of the energy gap points at a possibility of Co-oper pairs formation, although in terms of the free energy the preferred phase is the normal one. Thus, the supercritical region with a non-zero energy gap can be related to the emergence of the pseudogap. The phase transition to the superconducting

* e-mail address: Mateusz.Krzyzosiak@pwr.wroc.pl

state is found to be of the first order with a pronounced discontinuity in the entropy. The hysteresis phenomenon in the supercritical temperature range is considered in detail.

In order to study the response of the system to a weak magnetic field acting on both spin and charge degrees of freedom, the SFL model is extended by including the Pauli term and the superflow. In the extended model the eigenstates for the superconducting phase are found. The spin susceptibility and the magnetic field penetration depth are discussed. It is shown that the normal-phase spin susceptibility has a Curie-law cha-racter. Critical parameters such as the critical magnetic field and the critical current are studied employing the variational expansion of the free energy.

In: Superconductivity and Condensed Matter ... ISBN 978-1-61209-785-5
Editor: Collin J. Eckles

Chapter 58

FORMATION OF STRUCTURALLY DISORDERED YBCO UNDER MILD CONDITIONS BY THE USE OF CHEMICALLY ACTIVATED REAGENTS

***A. Knizhnik*[1] *and E. F. Makarov*[2]**
[1]Crown Center for Superconductivity and Department of Physics,
Technion-Israel Institute of Technology, Haifa 32000, Israel
[2]N. N. Semenov Institute of Chemical Physics,
Kosygin str 4, Moscow, Russia

RESEARCH SUMMARY

It was found that amorphous products of thermal decomposition of YBCO at low oxygen pressure have high chemical reactivity. In O_2 they can form YBCO again but already in more mild conditions (900°C, 45 h, without intermediate pulverizations). The lattice constants and the orthorhombic splitting of optimally oxygen doped YBCO coincide with the ones related to the structurally disordered modifications.

In: Superconductivity and Condensed Matter ...
Editor: Collin J. Eckles
ISBN 978-1-61209-785-5

Chapter 59

NANOSCALE INVESTIGATIONS ON MELT-TEXTURED AND SINGLE-CRYSTALLINE (Nd,Eu,Gd)$Ba_2Cu_3O_X$ ("NEG") HIGH-T_C SUPERCONDUCTORS

***M. R. Koblischka*[1], *P. Das*[1], *T. Wolf*[2], *M. Muralidhar*[3], *A. Koblischka-Veneva*[4] *and U. Hartmann*[1]**

[1]Institute of Experimental Physics, Saarland University, P.O.Box 151150, D-66041 Saarbrücken, Germany

[2]Institute of Solid State Physics, Forschungszentrum Karlsruhe, D-76021 Karlsruhe, Germany

[3]Superconductivity Research Lab, ISTEC, 1-10-13 Shinonome, Koto-ku, Tokyo 135-0062, Japan

[4]Institute of Functional Materials, Saarland University, P.O.Box 151150, D-66041 Saarbrücken, Germany

RESEARCH SUMMARY

Nanostripes are observed in light-rare-earth (LRE)-based ternary $(Nd_{0.33}Eu_{0.33}Gd_{0.33})Ba_2Cu_3O_X$ ("NEG") superconductors on both melt-textured and single crystalline samples. AFM/STM topography scans at ambient conditions on NEG melt-textured samples revealed the presence of nanostripes with a periodicity ranging between 15 nm and 60 nm and with a structure height of 0.3 – 10 nm. The recently developed single crystals of NEG exhibit nanostripes on the surface with a periodicity in the range 500 nm – 800 nm, and in the bulk nanostripes with a periodicity of about 20 to 40 nm. On the single crystals, low-temperature STM measurements were performed revealing the electronic differences within the nanostripes due to the presence of a LRE-rich phase. Employing AFM/STM topography scans together with electron backscatter diffraction (EBSD) analysis, we can determine the direction of the nanostripes with respect to the twin structure. Details of the STM topography measurements reveal that the nanostripes are composed by nanoclusters of the LRE-rich phase. The role of these nanostripes for flux pinning in these compounds is discussed.

In: Superconductivity and Condensed Matter … ISBN 978-1-61209-785-5
Editor: Collin J. Eckles

Chapter 60

RECENT ADVANCES OF NMR AND NQR STUDIES OF $YBA_2CU_3O_{7-\Delta}$ AND $YBA_2CU_4O_8$

Y. Itoh
Department of Chemistry, Graduate School of Science,
Kyoto University, 606-8502, Japan

RESEARCH SUMMARY

Nuclear magnetic resonance (NMR) and nuclear quadrupole resonance (NQR) techniques have supplied us with unique information on local incoherent phenomenon as well as low frequency magnetic response of electronic systems. Some of recent advances in the studies of high-T_c cuprate superconductors are based on site-selective measurements of the local electronic state and the local electron spin dynamics. High-T_c superconductors $YBa_2Cu_3O_{7-\delta}$(the optimal T_c = 93 K) and $YBa_2Cu_4O_8$ (T_c = 82 K) with double-layer CuO_2 planes have been studied by NMR and NQR tech-niques. From the intensive studies at the early stage, two-dimensional antiferromag-netic spin fluctuations and pseudo spin-gap in the magnetic excitation spectrum turned out to be key ingredients to characterize the novel electronic states, associated with the occurrence of superconductivity. Wipeout effect is a new key to characterize superconductor-insulator boundary in the lightly doped regime and impurity substi-tution effects. What happens near in-plane impurities and inside vortex cores in such novel electronic states? Nonmagnetic impurity Zn decreases T_c of the d-wave su-perconductivity because of the breakdown of Anderson theorem. From the careful NMR and NQR studies, it was concluded that the pseudo spin-gap in the normal state is robust for the substitution of nonmagnetic impurity Zn and magnetic impurity Ni. In the vicinity of a superconductor-insulator boundary induced by impurity, however, the pseudo spin-gap is suppressed. Collapse of low energy pseudo spin-gap due to the heavily substituted impurity Zn was also evidenced by neutron scattering measure-ments. For the impurity-substituted samples and in the vortex state of pure samples, the characteristic NMR and NQR spectra were observed. The local enhancement of mag-netic correlation near the impurity Zn and inside the vortex core was observed from the frequency distribution of nuclear spin-lattice relation time T_1. Scanning tunneling spectroscopy detects local density of states of electrons near Zn in the superconducting state. NMR and NQR experiments detect the local magnetism near Zn and inside the vortex cores.

In: Superconductivity and Condensed Matter …
Editor: Collin J. Eckles
ISBN 978-1-61209-785-5

Chapter 61

APPLICATION OF BULK SUPERCONDUCTORS TO SYNCHROTRON RADIATION SOURCES

T. Tanaka, R. Tsuru, T. Seike, H. Hirano, M. Morita, H. Teshima, S. Nariki, N. Sakai, I. Hirabayashi, M. Murakami and H. Kitamura
Precision Engineering Department, The University of Tokyo, Tokyo, Japan

RESEARCH SUMMARY

The synchrotron radiation (SR) is an electromagnetic wave emitted by a charged particle when deflected by a magnetic field, and has been utilized as a scientific probe in various research fields for many years. Nowadays more than 30 SR facilities are in operation all over the world. In these facilities, an electron beam is accelerated to an ultrarelativistic energy to produce intense SR, with which a variety of experiments and advanced researches are underway.

In order to enhance the SR intensity, a magnetic device is usually installed in a SR facility to generate a periodic magnetic field that makes an electron move along a peri-odic trajectory. Such a magnetic device is usually called a SR source. The SR emitted at each period is superimposed, resulting in much higher SR intensity. Although per-manent magnets (PMs) are usually used to generate the periodic magnetic field in the SR source because of their cost-effectiveness and availability, a large number of SR sources with superconducting magnets (SCMs) have been constructed for special uses, i.e., to obtain a strong magnetic field over 3 T, which cannot be achieved by using PMs alone. Most of these SR sources with SCMs are composed of superconducting coils made of low Tc (critical temperature) superconducting (LTS) material such as NbTi as in commercially available SCMs.

Another concepts of SR sources with SCMs have recently been proposed at the SPring-8, a SR facility in Japan, in which high Tc superconducting (HTS) bulk mag-nets are adopted instead of LTS coils. Because the nominal operating temperature can be much higher than liquid Helium, the limits on overall thermal budget will be much less stringent and operation will be much more feasible than the SR sources with LTS coils.

In order to utilize the HTS bulk magnets for the SR sources, it is necessary to establish a procedure for magnetizing an array of HTS bulk periodically so that it generates a periodic magnetic field. Two different schemes have been proposed so far. One is to make use of a mechanical movement of PM arrays in conventional SR sources, while the other is to apply an external field to an array of HTS bulk placed regularly. In the former procedure, the HTS bulk works as an additional magnet to enhance the magnetic field generated by the PM arrays, while the latter procedure takes full advantage of the capability of the HTS bulk. Experiments to demonstrate the two procedures described above have been performed, and promising results have been obtained.

In: Superconductivity and Condensed Matter …
Editor: Collin J. Eckles
ISBN 978-1-61209-785-5

Chapter 62

SQUID Based on YBCO Thin Film for Non-destructive Evaluation of Composite Materials

C. Bonavolontà
CNR-INFM "Coherentia", piazzale V.Tecchio 80,
80125 Napoli, Italy

Research Summary

The discovery of high temperature superconductors (HTS) in 1996 by Bednorz and Muller, led to an unprecedented explosion of research and development world-wide efforts, because of the significant potential for practical applications offered by these materials.

Then, the discovery of superconductivity in copper oxide perovskite, such as $YBa_2Cu_3O_{7-x}$, has opened a new era of research in superconducting materials. This class of materials not only show superconductivity higher than 90K, but also show properties that are different from classical superconductors. This offers a great challenge to understanding the basic phenomenon that causes superconductivity in these materials and to developing new fabrication methods for thin film and device. One of the most important device made of YBCO film is the Superconducting Quantum Interference Device (SQUID), an extremely sensitive magnetic flux-to-voltage transducer, used to measure extremely tiny magnetic fields.

Typical applications areas of HTS SQUID included metrology and electronic instrumentation, medical diagnostics, electronic warfare (radar, EMC, magnetic anomaly detection), non-destructive materials evaluation (NDE-NDT), geological prospecting and data processing.

This work has the aim to show the potentiality and the performance of devices made of YBCO thin films, such as SQUID, in the field of material science. In this chapter an overview of the SQUID NDE application with a particular focus on the analysis of non metallic composite materials is presented. The chapter begins with a description of the fundamental properties and the performance that characterize the SQUID based on YBCO thin films. The other sections of the chapter will be dedicated to the several NDE applications of the SQUID, focusing the attention on the testing of carbon fiber reinforced composite, widely used in the aerospace and aircraft industry.

In: Superconductivity and Condensed Matter … ISBN 978-1-61209-785-5
Editor: Collin J. Eckles

Chapter 63

NEW APPLICATIONS FOR PINNING EFFECT

Mojtaba Ghodsi, Toshiyuki Ueno and Toshiro Higuchi
Precision Engineering Department, The University of Tokyo, Tokyo, Japan

RESEARCH SUMMARY

Discovery of new oxides of superconductors (YBaCuO, DyBaCuO, GdBa-CuO,...) opened the large opportunities for industries. Within the compass of this chapter, we only try to highlight the potential of pinning effect occurred in type-II of supercon-ductors. For examples two new applications of pinning effect which are (i)passive levitation of soft magnetic material by bulk superconductor and (ii)zero-power cryo-genic actuator will be described.

In: Superconductivity and Condensed Matter ...
Editor: Collin J. Eckles

ISBN 978-1-61209-785-5

Chapter 64

PROCESSING OF HIGH-PERFORMANCE RE-BA-CU-O BULK SUPERCONDUCTORS

Shinya Nariki[1], Naomichi Sakai[1], Masato Murakami[2] and Izumi Hirabayashi[1]

[1]Superconductivity Research Laboratory, ISTEC, Tokyo 135-0062 Japan
[2]Shibaura Institute of Technology, Tokyo 135-8548 Japan

RESEARCH SUMMARY

Large grain RE-Ba-Cu-O (RE: rare earth elements) bulk superconductors can trap large magnetic fields and thus can function as a strong quasi-permanent magnet. For the enhancement of trapped magnetic field of bulk superconductors, it is important to increase the critical current density (J_c) and/or the size of single-domain. In this chapter, the performance of recently developed large-grain RE-Ba-Cu-O superconductors is introduced. The J_c values of RE-Ba-Cu-O superconductors are strongly dependent on the microstructure. The introduction of excess fine RE211 contributes to the enhancement of J_c. In the case of large single-grain Gd-Ba-Cu-O bulk sample of 65 mm diameter with fine Gd211 particles, the maximum trapped magnetic field of 3.05 T was recorded at 77 K. Furthermore, the employment of ball-milled ultra-fine RE211 powder as a precursor material is very effective in improving J_c, since RE211 inclusions in the melt-textured samples could be drastically refined with decreasing the particle size of RE211 starting powder. In (Gd, Y)-Ba-Cu-O bulk sample, the J_c value reached an extremely high value of 380 kA/cm^2 at 77 K in self-field. We also introduced the superconducting properties of Y- and Gd-Ba-Cu-O bulk superconductors with $BaCeO_3$ addition as a novel flux pinning center.

In: Superconductivity and Condensed Matter ... ISBN 978-1-61209-785-5
Editor: Collin J. Eckles

Chapter 65

THE EFFECT OF DEFORMATION TEMPERATURE ON THE MICROSTRUCTURE AND TEXTURE IN $YBa_2Cu_3O_{7-x}$ CERAMICS PROCESSED BY TORSION UNDER PRESSURE

M. F. Imayev, D. B. Kabirova and A. V. Dementyev
Institute for Metals Superplasticity Problems, Russian Academy of Sciences,
Khalturina 39, Ufa 450001, Russia

RESEARCH SUMMARY

The effect of deformation by torsion under quasi-hydrostatic pressure in the temperature range 900-1035°C on the evolution of the microstructure and texture of $YBa_2Cu_3O_{7-x}$ (Y123) ceramics was studied. It was established that dependencies of parameters of grains (mean length and thickness) and the degree of c-axis tilt alignment parallel to the direction of compression on deformation temperature are similar and correlate with the temperature ranges of liquid emergence at grain boundaries. Such correlated temperature behavior is caused by the fact that both the parameters of grains and the degree of basal plane texture depend on the amount of liquid at grain boundaries. There are four characteristic temperature ranges of the correlated change of parameters of grains and degree of basal texture. In the range 900-935°C the sizes of grains and degree of texture grow. The aspect ratio of grains in this temperature range also grows from 2 to 3.5. Such behavior is caused by appearance of an eutectic liquid at grain boundaries in the temperature range 900-930°C. In the range 935-1000°C the grain's parameters and degree of basal texture remain at a fixed level that is connected with the constant amount of liquid in the given temperature range. Above 1000°C the grain sizes and basal texture grow again due to the appearance of a new portion of liquid as a result of incongruent melting onset of Y123 phase. The grain sizes and basal texture reach their maximum values at 1020°C and further decrease. The regularities of microstructure change and mechanism of texture formation during high-temperature deformation of the Y123 superconductor are discussed.

In: Superconductivity and Condensed Matter ...
Editor: Collin J. Eckles
ISBN 978-1-61209-785-5

Chapter 66

EMBEDDING NANOPARTICLES AS FLUX PINNING SITES INTO MELT-TEXTURED YBCO BULKS – AN EBSD STUDY

A. Koblischka-Veneva[1], M. R. Koblischka[2], N. Hari Babu[3] and D. A. Cardwell[3]

[1]Institute of Functional Materials, Saarland University, Saarbrücken, Germany
[2]Institute of Experimental Physics, Saarland University, Saarbrücken, Germany
[3]IRC in Superconductivity, University of Cambridge, Cambridge, UK

RESEARCH SUMMARY

Microstructures and crystallographic orientations of melt-textured YBCO samples with embedded nano-sized particles of Y_2BaCuO_5 (Y-211) and $Y_2Ba_4CuMO_x$ (M-2411 with M = U, Nb, Zr, Ag) are analysed by means of electron backscatter diffraction (EBSD). With the recent developments of the EBSD technique, we can directly measure the crystallographic orientation of the embedded nanoparticles, employing a spatial resolution of about 40 nm. The high image quality of the Kikuchi patterns allows true three-phase (YBCO, Y-211 and M-2411) scans to be performed. The resulting texture of the superconducting YBCO matrix is found not to be disturbed by the presence of these nanoparticles. The Y-211 particles do not exhibit any preferred orientation, even if their size is considerably reduced to the 100 nm range. The size reduction reduces, however, the negative influence of the Y-211 particles on the YBCO matrix. U-2411 particles, which are formed during the processing stage, do not show any orientation, and with increasing concentration, some texture develops. In contrast to this, embedded Zr-2411 particles are fully oriented in (0 0 1) orientation according to the surrounding superconducting matrix. By adding oxides like Nb_2O_5 and Y_2O_3, the respective nanoparticles are created during the reaction, but as a consequence, there is no significant texture of these nanoparticles. If, in contrast, the $Y_2Ba_4CuMO_x$ particles are added as previously prepared nanopowders, they are found to exhibit only one dominant orientation analoguous to the surrounding YBCO matrix. Therefore, the EBSD analysis clearly shows the pathway for further improvement of the performance of bulk high-T_c superconductors.

In: Superconductivity and Condensed Matter ...
Editor: Collin J. Eckles

ISBN 978-1-61209-785-5

Chapter 67

COMPARATIVE STUDY OF STATISTICS OF COOPER'S ELECTRON PAIRS IN LOW-TEMPERATURE SUPERCONDUCTORS AND COUPLED HOLES PAIRS IN HIGH T_C CERAMICS

I. G. Kaplan** and O. Navarro

Instituto de Investigaciones en Materiales, Universidad Nacional Autonoma de Mexico, Apartado Postal 70-360, 04510, Mexico

RESEARCH SUMMARY

It is well known that the system of Cooper's pair is described by boson symmetric wave functions, but Cooper's pair operators are bosons only when the moments **k** are different and they are fermions for equal **k**. The analysis of trilinear commutation relations for the Cooper pair (pairon) operators reveals that they correspond to the modified parafermi statistics of rank p=1. Two different expressions for the Cooper pair number operator are presented. We demonstrate that the calculations with a Hamiltonian expressed via pairon operators is more convenient using the commutation properties of these operators without presenting them as a product of fermion operators. This allows to study problems in which the interactions between Cooper's pairs are also included. The problem with two interacting Cooper's pairs is resolved and its generalization in the case of large systems is discussed.

It is shown that in site representation, the hole-pair operators obey the same commutation relations (paulion) as the Cooper pair operators in impulse representation, although the latter describe delocalized quasiparticles. In quasi-impulse representation, the hole-pair operators are also delocalized and their exact commutation relations correspond to a modified parafermi statistics of rank M($M$is the number of sites in a "superlattice" formed by the centers of mass of each hole pair). From this follows that one state can be occupied by up to Mpairs. Even in the absence of dynamic interaction, the system of hole pairs is characterized by some immanent interaction, named after Dyson as kinematic interaction. This interaction appears

* E-mail address: kaplan@zinalco.iimatercu.unam.mx

because of the deviation of the quasiparticle statistics from the Bose (Fermi) statistics and its magnitude depends on the concentration of hole pairs. In spite of the non-bosonic behavior, there is no statistical prohibition on the Bose-Einstein condensation of coupled hole pairs.

In: Superconductivity and Condensed Matter …
Editor: Collin J. Eckles
ISBN 978-1-61209-785-5

Chapter 68

SUPERCONDUCTIVITY IN HIGHLY CORRELATED SYSTEMS

Hideo Aoki
Department of Physics, University of Tokyo, Hongo, Tokyo 113-0033, Japan

RESEARCH SUMMARY

The problem we would like to first address here is the opposite limit of short-range repulsion, as appropriate for strongly-correlated systems such as transition metal oxides. There, the pairing interaction is mainly spin-fluctuation mediated. The most widely used model is the Hubbard model having the on-site repulsion, U. If superconductivity can arise in the one-band Hubbard model, the simplest possible (hence generic) model for repulsively correlated electron systems, the interest is very fundamental indeed, which should serve as a guiding principle if one wishes to have even higher TC superconductivity. This chapter will look at how this kind of anisotropic pairing superconductivity can be ``optimised", namely how we can make TC higher [1,2]. We first note that ``TC is very low in the electron mechanism" in that TC is usually two orders of magnitude lower than the electronic energy. We propose how this can be overcome. We shall also compare the correlated electron systems as exemplified by high-Tc cuprate with another prototype, the fractional quantum Hall system.

In: Superconductivity and Condensed Matter ...
Editor: Collin J. Eckles
ISBN 978-1-61209-785-5

Chapter 69

PLASMON EXCHANGE MODEL FOR SUPERCONDUCTIVITY IN CARBON NANOTUBES

S. M. Bose **[*]** ***and S. Gayen***
Department of Physics, Drexel University, Philadelphia, PA 19104, US

RESEARCH SUMMARY

Recent experiments have shown that a single-wall zigzag carbon nanotube can become superconducting at around 15 K. Calculations using the traditional phonon exchange model, however, give a superconducting transition temperature T_C less than 1K. To explain the observed higher critical temperature the plasmon exchange mechanism for superconductivity in nanotubes is explored in this paper. Calculation of the effective interaction between electrons in a nanotube mediated by plasmon exchange shows that this interaction can become attractive. This attractive interaction is then used in the modified Eliashberg theory for strong coupling superconductors to calculate the critical temperature T_C in a nanotube. T_C is found to be sensitively dependent on the dielectric constant of the medium, the effective mass of the electrons and the radius of the nanotube. These results can explain the observed T_C in a nanotube.

[*] E-mail address: bose@drexel.edu Telephone: +1-215-895-2718.

In: Superconductivity and Condensed Matter ...
Editor: Collin J. Eckles
ISBN 978-1-61209-785-5

Chapter 70

THERMODYNAMIC PROPERTIES OF POINT NODESUPERCONDUCTORS: BOROCARBIDES

Heesang Kim
Department of Physics, University at Buffalo, State University of New York, Buffalo, NY 14260, US
Department of Physics, Soongsil University, Seoul 156-743, Korea

RESEARCH SUMMARY

We investigate interesting properties of a point node superconductor within a framework of the weak coupling theory. We mainly focus on the effect of non-magnetic impurities on the density of states and the transition temperatures. The results can be applied to the borocarbide superconductors. Our model calculation suggests that the assumption of the point node order parameter in the borocarbides reconciles the current controversies in experimental results.

In: Superconductivity and Condensed Matter ... ISBN 978-1-61209-785-5
Editor: Collin J. Eckles

Chapter 71

COMPELLING EVIDENCE FOR HIGH-TEMPERATURE SUPERCONDUCTIVITY IN CARBON NANOTUBES

Guo-meng Zhao*
Department of Physics and Astronomy,California State University at Los Angeles, Los Angeles, CA 90032, US

RESEARCH SUMMARY

In this article, we provide compelling evidence for high temperature supercon-ductivity in carbon nanotubes. Although the superconducting transition temperatures can vary from 0.44 K to 750 K, the resistive transitions in these superconducting carbon nanotubes are in quantitative agreement with the well-established Langer-Ambegaokar-McCumber-Halperin (LAMH) theory for the resistive transition in quasi-one-dimensional superconductors. The Meissner effect is clearly seen in aligned and physically separated multi-walled nanotubes up to room temperature. At low tem-peratures, the directionally averaged diamagnetic susceptibility for physically coupled multi-walled nanotubes increases by a factor of 4.3 compared with that for physically separated tubes, which is consistent with the enhancement of the Meissner effect due to the Josephson coupling among the superconducting tubes. Because of a finite num-ber of transverse conduction channels, the four-probe resistance will never go to zero in the superconducting state. Nonetheless, the on-tube resistance at room temperature has been found to be indistinguishable from zero for many individual multi-walled nanotubes. The very small but finite room-temperature on-tube resistance is consistent with quantum phase slips due to the finite number of transverse channels. Further, Raman data and single-particle tunneling spectra consistently show the existence of single particle excitation gaps of larger than 100 meV. We suggest microscopic pairing mechanism for high-temperature superconductivity in carbon nanotubes.

---

* E-mail address: gzhao2@calstatela.edu

In: Superconductivity and Condensed Matter ... ISBN 978-1-61209-785-5
Editor: Collin J. Eckles

Chapter 72

MAGNETISM AND SUPERCONDUCTIVITY IN QUARK MATTER

Toshitaka Tatsumi[1], Eiji Nakano[2] and Kanabu Nawa[1]

[1]Department of Physics, Kyoto University, Kyoto 606-8502, Japan

[2]Yukawa Institute for Theoretical Physics, Kyoto University, Kyoto 606-8502, Japan

RESEARCH SUMMARY

Magnetic properties of quark matter and its relation to the microscopic origin of the magnetic field observed in compact stars are studied. Spontaneous spin polariza-tion appears in high-density region due to the Fock exchange term, which may provide a scenario for the behaviors of magnetars. On the other hand, quark matter becomes unstable to form spin density wave in the moderate density region, where restoration of chiral symmetry plays an important role. Coexistence of magnetism and color su-perconductivity is also discussed.

In: Superconductivity and Condensed Matter …
Editor: Collin J. Eckles
ISBN 978-1-61209-785-5

Chapter 73

SUPERCONDUCTIVITY AND MAGNETISM IN $RuR_{2-x}Ce_xSr_2Cu_2O_{10-\Delta}$ (R=EU AND GD)

I. Felner
Racah Institute of Physics, The Hebrew University, Jerusalem, 91904, Israel

RESEARCH SUMMARY

The magneto-superconducting $RuR_{2-x}Ce_xSr_2Cu_2O_{10-\delta}$ (R=Eu and Gd, Ru-1222) materials are the first Cu-O based system in which bulk superconductivity (SC) in the CuO2 planes (below T_C = 32-50 K) and *weak*-ferromagnetism (W-FM) in the Ru sublattice, coexists. The hole doping of the Cu-O planes, which results in metallic behavior and SC, can be optimized with appropriate variation of the R/Ce ratio and/or the oxygen concentration. SC survives because the Ru magnetic moments probably align in the basal planes, which are practically decoupled from the CuO_2 planes.

Two magnetic transitions are observed in Ru-1222, at T_M~160 K and T_{2M} ~80 K. The higher ordering temperature at T_M, does not change with x and one of the disputed question is its *origin*, namely, whether it is intrinsic or not. Below T_{2M} the Ru moments are W-FM ordered and wide ferromagnetic-like hysteresis loops are obtained. They become narrow and disappear at ~ 60-70 K. Above T_{M2} (around 120 K), (i) a small peak is observed in both dc and ac susceptibility curves and (ii) narrow hysteresis loops reappear with a peak in the coercive fields around 120 K. The accumulated results for the higher magnetic phase are interpreted by two scenarios (a) The two observed phenomena are related to intrinsic reduction of pure Ru^{5+} ions (in x=1) to Ru^{4+} caused by the variation of the Ce^{4+} content. This may form Ru^{4+} rich clusters, in which the Ru^{4+}-Ru^{4+} exchange interactions are stronger than the Ru^{5+}-Ru^{5+} interactions. (b) It is attributed to the presence of nano-particles of a foreign minor extra Ru^{4+} magnetic phase such as Sr-Cu-Ru-O_3. The second scenario is supported by Mössbauer spectroscopy of ^{57}Fe doped materials.

In: Superconductivity and Condensed Matter ...
Editor: Collin J. Eckles

ISBN 978-1-61209-785-5

Chapter 74

APPLICATIONS OF EXTENDED CRITICAL STATE MODEL TO HIGH T_C SUPERCONDUCTORS

Yoshihide Kimishima
Department of Physics, Division of Physics, Electrical and Computer Engineering,
Graduate School of Engineering, Yokohama National University
Tokiwadai 19-5, Hodogaya-ku, Yokohama 240-8501, Japan

RESEARCH SUMMARY

Analytical expressions of magnetization M and critical current density J_c by an extended critical state model. The formula of *Kimishima model* was one of the universal functions interconnecting the Bean and Anderson-Kim models. The second formula of *Exponential Kimishima model* was derived by the extension of *Kimishima model*, considering the exponential decay of J_c near the irreversible field H_{irr}. These theoretical magnetization curves were well fitted to the experimental ones. Estimated magnetic parameters were important to explain the temperature dependences of J_c and the pinning properties of high T_c superconductors.

In: Superconductivity and Condensed Matter …
Editor: Collin J. Eckles
ISBN 978-1-61209-785-5

Chapter 75

SPONTANOUS VORTEX PHASE, PINNING AND TRANSPORT IN HETEROGENOUS FERROMAGNETIC-SUPERCONDUCTING SYSTEMS

M. Amin Kayali
Human Neuroimaging Laboratory, Baylor College of Medicine,
Houston, TX 77030, US

RESEARCH SUMMARY

The past few years have witnessed an increasing interest in studying the static and dynamic phases of heterogeneous ferromagnetic-superconducting systems. Typically, a heterogeneous ferromagnetic-superconducting system consists of a superconductor placed in close proximity with a ferromagnet separated by a layer of insulator oxide. The layer of insulator oxide suppresses proximity effects and spin diffusion hence al-lowing for the coexistence of two competing order parameters, a feature that is special to these systems. The strong interaction between the ferromagnet and the superconduc-tor gives raise to strong vortex pinning force which enhances the critical current of the superconductor. In this chapter, I will review the important theoretical developments in this area of research. The first part of this review will focus on the static properties of these systems; therefore, the general formulation of Londons theory for these systems will be presented. Afterward, I will apply the general theory to study few specific sys-tems with emphasis on the appearance of spontaneous vortex phase and the important role played by the geometry and the magnetization distribution of the ferromagnet on the static and dynamic properties of the superconductor. In the second part, I will con-centrate on the dynamic of vortices and antivortices in these systems. Therefore, I will review the anisotropic transport properties of a ferromagnetic-superconducting bilayer then I will study the dynamical properties of a system consisting of a superconducting thin film pierced by a square array of ferromagnetic columnar defects.

In: Superconductivity and Condensed Matter …
Editor: Collin J. Eckles
ISBN 978-1-61209-785-5

Chapter 76

SCANNING SQUID MAGNETIC IMAGING FOR NONDESTRUCTIVE EVALUATION: FROM MICRON-SCALE RESEARCH TO PRACTICAL LARGE-SCALE APPLICATION

K. Isawa
R & D Center, Tohoku Electric Power Co., Inc.
2-1 Nakayama 7-chome, Aoba-ku, Sendai, 981-0952 Japan

RESEARCH SUMMARY

Superconducting magnetic sensor SQUID is the most sensitive magnetic sensing element known that has high spatial resolution. Accordingly, scanning SQUID magnetic imaging is currently being developed for applications to NDE (nondestructive evaluation). Laboratory-scale scanning SQUID microscopes (SSM) are useful for academic research as well as for industrial applications that require high-resolution measurements. This paper firstly reviews current knowledge of (a) temperature dependence of local magnetic distribution in high-Tc superconductors by SSM, and (b) high-resolution measurements using SSM on stress-induced magnetism in austenitic. For practical applications, a cost-effective NDE method that is highly accurate is desired. SQUID-NDE techniques are also reviewed here because they were recently proved capable of detecting damage such as fatigue, deformation, and cracks in materials, especially metals. Considering the demand in industry, conventional stationary SQUID-NDE methods are practically useless for inspecting large structures or material components. To answer this demand, an advanced SQUID-NDE system, called a *traveling SQUID-NDE*, was recently developed. This traveling SQUID-NDE can continuously "travel" over the surface of an object during testing without the need for magnetic shielding. And finally, this system is also reviewed here.

In: Superconductivity and Condensed Matter … ISBN 978-1-61209-785-5
Editor: Collin J. Eckles

Chapter 77

CHEMICAL AND MACROSCOPIC PHYSICAL PROPERTIES OF THE SODIUM COBALT OXYHYDRATE SUPERCONDUCTOR

***Hiroya Sakurai*[1,2], *Kazunori Takada*[3] and *Eiji Takayama-Muromachi*[1]**

[1]Superconducting Materials Center, National Institute for Materials Science, Namiki 1-1, Tsukuba, Ibaraki 305-0044, Japan

[2]International Center for Young Scientists, National Institute for Materials Science, Namiki 1-1, Tsukuba, Ibaraki 305-0044, Japan

[3]Advanced Materials Laboratory, National Institute for Materials Science, Namiki 1-1, Tsukuba, Ibaraki 305-0044, Japan

RESEARCH SUMMARY

This report reviews chemical and macroscopic physical properties of the bilayer sodium cobalt oxyhydrate (BLH) superconductor and its related compounds, focusing on the recent development of research. The exact composition of BLH is $Na_x(H3O)_zCoO2{\cdot}y'H_2O$ and it has three adjustable compositional parameters: Na content, x, H_3O content, z, and H_2O content, y'. The former two parameters greatly affect superconductivity and related properties of the system. With adjusting them, a magnetically ordered phase appears in the immediate vicinity of the superconducting phases in a phase diagram, suggesting a certain magnetic origin for superconductivity. Regarding the band structure of BLH, hole pockets near the K points look crucially important relating to the mechanism of superconductivity. Furthermore, there is a possibility that two-gap superconductivity is realized in the present system corresponding to the hole pockets near the K points and a large cylindrical Fermi surface around the Γ point. At the end of the paper, the mother oxide of γ-Na_xCoO_2 is discussed. Its rich physical properties originate from strong electron-electron correlation in the CoO_2 layer and seem to relate to the unconventional superconductivity in BLH.

In: Superconductivity and Condensed Matter … ISBN 978-1-61209-785-5
Editor: Collin J. Eckles

Chapter 78

FABRICATION AND TRANSPORT PROPERTY OF DIRECTIONALLY SOLIDIFIED RE123

Yuichi Nakamura, Ryoji Inada and Akio Oota
Toyohashi University of Technology, Toyohashi, Aichi 441-8580, Japan

RESEARCH SUMMARY

$REBa_2Cu_3O_y$ (RE123: RE=Y, Sm, Gd etc.) superconductive oxides have potential for many applications since they exhibit a high critical current density (J_c) under high magnetic fields at 77 K. In this paper, a directional solidification of RE123 (RE=Y and Sm) samples was carried out. The effect of fabrication conditions on the microstructure and transport J_c properties of Y123 fibers were reported, and the continuous growth condition of Sm123 are also investigated. The transport J_c property was measured by the standard d.c. 4 probe method. The continuous growth of the 0.5 mmϕ Y123 fiber was achieved in the samples grown at high pulling rates up to 10 mm/h while the usual Y123 rods of 2 mmϕ could grow continuously up to several mm/h. The size of Y211 particles trapped in Y123 was found to decrease with increasing pulling rate. The Sm123 can continuously grow at higher pulling rates than Y123 for both fibers and rods. The transport J_c properties of Y123 fibers showed the maximum in the sample pulled at 6 mm/h; the best I_c and corresponding J_c values achieved were 134 A and 1□ 10^5 A/cm^2 at 77K self-field, respectively. This J_c value is the best transport J_c of the melt grown samples to date and is comparable to the best J_c value of small Y123 bulks. This high J_c value may attributed to the well-controlled microstructure and crystal orientation of the sample fabricated at high pulling rate. That is, the preferential crystal orientation as well as fine Y211 particles is thought to result this high transport J_c of the fiber sample. On the other hand, the enlargement of continuous growth condition for the fiber samples could be fundamentally explained by the difference in effective nucleation rate, which was defined as the product of nucleation frequency and undercooled volume in the liquid. Because of the difference in the cross sectional area, the small nucleation volume of thin fiber leads the reduction of the nucleation rate, and may result in the increase of the continuous growth rate. This difference in nucleation rate is, however, found not to be enough to explain the destruction of continuous growth, and the additional factor such as compositional shift at the interface is needed. To think of a reason causing the compositional shift, a non-steady state growth model is proposed, and the effect of process parameters and growth conditions on the compositional shift is discussed.

In: Superconductivity and Condensed Matter ... ISBN 978-1-61209-785-5
Editor: Collin J. Eckles

Chapter 79

A MOLECULAR DYNAMICS DESCRIPTION OF CLUSTERS IN STRONG FIELDS*

M. Belkacem,[1] F. Megi,[1] E. Suraud,[1] P.-G. Reinhard[2] and G. Zwicknagel[2]

[1]Laboratoirede Physique Th´eorique Universit´e P. Sabatier, 118 Rtede Narbonne F-31062 Toulousecedex 4, France

[2]Institut f¨ur Theoretische Physik Universit¨at Erlangen Staudtstrasse 7, D-91058 Erlangen, Germany

RESEARCH SUMMARY

We present here a particularly simple MD model, where core and valence electrons are treated at the same level of classically Coulomb interacting particles, while a minimum of quantum information is introduced by associating a finite width to each electron.

* Paper Presented at the 29th International Workshop on Condensed Matter Theories, Kizu, Kyoto, Japan, September 13-17, 2005.

In: Superconductivity and Condensed Matter ... ISBN 978-1-61209-785-5
Editor: Collin J. Eckles

Chapter 80

EXTENDED UNIVERSALITY IN A STATISTICAL MECHANICAL MODEL

Carlos Wexler, Cintia M. Lapilli and Peter Pfeifer
Department of Physics and Astronomy, University of Missouri–Columbia,
Columbia, Missouri, 65211, US

RESEARCH SUMMARY

In this paper, we report a different type of strong universality. We present the surprising result that, in a specific family of systems, different members behave identically both near and away from critical points-we refer to this as extended universality, if the temperature and a parameter p, describing the interaction between neighboring molecules, exceed a certain value. In this regime, the thermodynamic observables collapse, in the sense that their values are identical for different values of p. No thermodynamic measurements in this regime can reveal the details of the interaction in the Hamiltonian. This sets limits on the microscopic information that can be extracted from macroscopic measurements. As the collapse maps Hamiltonians with different symmetries onto one and the same thermodynamic state, the system exhibits a symmetry not present in the Hamiltonian. The added symmetry at high temperature is the counterpart of broken symmetry at low temperature. To the best of our knowledge, no such collapse of thermodynamic observables has been observed before.

In: Superconductivity and Condensed Matter ...
Editor: Collin J. Eckles
ISBN 978-1-61209-785-5

Chapter 81

METAL L-INSULATOR PHENOMENA IN 2D: A UNIFIED SCALING PICTURE*

D.J.W. Geldart
Department of Physics, Dalhousie University Halifax NS, Canada
School of Physics, University of New South Wales Sydney 2052, Australia
Dipartimentodi Fisica, Universit`adi Camerino Camerino (MC) Italy
D. Nelson
Dipartimentodi Fisica, Universit`adi Camerino Camerino (MC) Italy
School of Physics, University of New South Wales Sydney 2052, Australia

RESEARCH SUMMARY

In this paper we discuss the 2D metal-insulator transition phenomenon, which is one of these new phenomena and which suggests in such devices that increasing the electron density causes a transition in the ground state from an insulating ground state to a metallic ground state. Associated with this phenomenon is some evidence for an apparent new quantum critical point that would separate the insulating and metallic ground states. The observed scaling of the temperature dependent resistivity with the deviation of electron density from a critical density in Siand GaAs suggests the existence of a quantum critical transition at critical density n_c in the $T \to 0$ limit. In spite of a great deal of experimental and theoretical effort the existence of this phenomenon remains controversial, and we discuss here some new ideas for resolving the issue of whether or not there exists metal-insulator transition phenomenon in 2D.

* Paper Presented at the 29th International Workshop on Condensed Matter Theories, Kizu, Kyoto, Japan, September 13-17, 2005.

In: Superconductivity and Condensed Matter … ISBN 978-1-61209-785-5
Editor: Collin J. Eckles

Chapter 82

PHASE TRANSITIONS AND HYSTERESIS IN A SYSTEM OF TWO COUPLED MAGNETIC NANOPARTICLES*

Karl E. Kürten
Institute for Experimental Physics, University of Vienna 5, Boltzmanngasse, A- 1090 Vienna, Austria and Department of Physics, Loughborough University, Loughborough, LE113TU, UK

RESEARCH SUMMARY

We studied dynamical and static properties of two interacting magnetic particles subjected to exchange interaction J, anisotropy parameter K and an external magnetic field H with an arbitrary angle β with respect to the easy axis. The well-known Stoner Wohlfahrt model for a single magnetic particle was generalized to model the critical behavior of two anti-ferromagnetically coupled ferromagnets in order to have a description for an antiferro-magnet. Moreover, our generalized Stoner "asteroid" is easily accessible for experimental studies. We further present a complete theoretical study of magnetic phase diagrams as a function of the field for arbitrary strength of the anisotropy. We classify all possible magnetic hysteresis loops and show the dependence of the corresponding magnetic moments.

* Paper Presented at the 29th International Workshop on Condensed Matter Theories, Kizu, Kyoto, Japan, September 13-17, 2005.

In: Superconductivity and Condensed Matter ... ISBN 978-1-61209-785-5
Editor: Collin J. Eckles

Chapter 83

EFFECT OF APICAL OXYGEN ORDERING ON T_C OF CUPRATE SUPERCONDUCTORS*

***H. Yang, Q. Q. Liu, F. Y. Li, C. Q. Jin and R. C. Yu*[1]**
Key Laboratory of Extreme Conditions Physics,
Beijing National Laboratory for Condensed Matter Physics,
Institute of Physics, P. O. Box 603, Beijing 100080, P. R. China

RESEARCH SUMMARY

Chemical disorder introduced into the charge reservoir blocks by doping has been shown to be one of the parameters influencing the superconducting transition temperature (Tc) of high-Tc cuprate superconductors (HTSs). We address the question of whether the Tc is susceptible to the ordering of dopant atoms. The answer is found by studying the Sr2CuO3+δ superconducting system with K2NiF4-type structure in which oxygen atoms only partially occupy the apical sites next to the CuO2 planes and act as hole dopants. The numerous Tcs appearing in this system are revealed to arise from different modulated phases that are formed just by the ordering of apical oxygen, and each of the superconducting modulated phases is associated with a distinct type of the ordering. The superconductivity differences for the modulated phases are revealed to result, mainly, from the ordering of apical oxygen.

* Paper Presented at the 29th International Workshop on Condensed Matter Theories, Kizu, Kyoto, Japan, September 13-17, 2005.
[1] E-mail address: rcyu@aphy.iphy.ac.cn (Corresponding author).

In: Superconductivity and Condensed Matter ...
Editor: Collin J. Eckles

ISBN 978-1-61209-785-5

Chapter 84

GEOMETRIC EFFECTS AND MAGNETIC PHASES IN QUANTUM RINGS*

S. Janecek, M. Aichinger, E.R ¨as¨ anen and E. Krotscheck
Institut f¨ur Theoretische Physik Johannes Kepler Universit¨at,
A- 4040 Linz, Austria

S.A. Chin
Department of Physics Texas A&M University
College Station, TX 77843-4242, US

RESEARCH SUMMARY

Our goal in this paper is to connect the above-mentioned experimental results on a general level by computing the energetics of circular and square two-dimensional (2D) quantum rings containing up to $N \sim 20$ electrons with a considerable Coulomb interaction. We calculate the properties of these systems within spin density functional theory (SDFT). The Kohn-Sham (KS) wave functions are computed by applying a split operator technique that allows us to exactly decompose the kinetic part of the Hamiltonian including a uniform external magnetic field. Numerical results for the relevant observable quantities such as chemical potentials and magnetization a represented in section 4.

* Paper Presented at the 29th International Workshop on Condensed Matter Theories, Kizu, Kyoto, Japan, September 13-17, 2005.

In: Superconductivity and Condensed Matter … ISBN 978-1-61209-785-5
Editor: Collin J. Eckles

Chapter 85

RECENT ADVANCES IN THE ELECTRONIC AND VIBRATIONAL STUDIES OF CARBON NANOTUBES USING FULLERENES AS LOCAL PROBES *

Ferenc Simon †
Budapest University of Technology and Economics, Institute of Physics and Condensed Matter Research Group of the Hungarian Academyof Sciences, P.O. Box 91, H-1521, Budapest, Hungary
Institut f¨ur Materialphysik, Universit¨at WienStrudlhofgasse 4, A-1090 Wien, Austria

RESEARCH SUMMARY

The discovery of fullerenes encapsulated inside single-wall carbon nanotubes (SWCNTs) by Smith *et al.* [Nature 396, 323 (1998)], provides a compelling oppor-tunity to study the nanotubes using fullerenes as local probes. It is reviewed herein, how fullerenes and magnetic fullerenes such as N@C_{60} and $C_{59}N$, and ^{13}C enriched fullerenes can be used to yield unprecedented insight into the nanotube electronic, vi-brational, and optical properties and into the growth of SWCNTs. Encapsulated C_{60} fullerenes are transformed to inner tubes by high temperature annealing. The unique, low defect concentration and well shielded nature of inner tubes makes them ideal to study the effect of diameter dependent treatments such as opening and closing of the tubes. Electron-spin resonance studies on the encapsulated magnetic fullerenes indi-cates a metallic-like finite density of states for an ensemble SWCNT sample, which contradicts the current view of SWCNTs as containing separate metallic and semicon-ducting nanotubes. The linear spin chains of the encapsulated magnetic fullerenes is suggested as a building element for quantum information processing. The possibility to grow inner tubes from encapsulated fullerenes allows the isotope selective enrich-ment of the inner tubes if the source fullerene is ^{13}C enriched. Nuclear magnetic resonance on such samples

* Paper Presented at the 29th International Workshop on Condensed Matter Theories, Kizu, Kyoto, Japan, September 13-17, 2005.
† E-mail address: ferenc.simon@univie.ac.at

evidences the presence of a low energy spin-gap which is thought to open due to a Peierls transition.

In: Superconductivity and Condensed Matter …
Editor: Collin J. Eckles
ISBN 978-1-61209-785-5

Chapter 86

DYNAMICAL SIMULATION OF NUCLEAR "PASTA": SOFT CONDENSED MATTER IN DENSE STARS

***Gentaro Watanabe*[1,2] *and Hidetaka Sonoda*[2,3]**
[1]NORDITA, Blegdamsvej 17, DK-2100 Copenhagen Ø, Denmark
[2]The Institute of Chemical and Physical Research (RIKEN)Saitama 351-0198, Japan[c]Department of Physics, University of Tokyo, Tokyo 113-0033, Japan

RESEARCH SUMMARY

More than twenty years ago, it was predicted that nuclei can adopt interesting shapes, such as rods or slabs, etc., in the cores of supernovae and the crusts of neutron stars. These non-spherical nuclei are referred to as nuclear "pasta."

In recent years, we have been studying the dynamics of the pasta phases using a method called quantum molecular dynamics (QMD) and have opened up a new as-pect of study for this system. Our findings include: dynamical formation of the pasta phases by cooling down the hot uniform nuclear matter; phase diagrams in the density versus temperature plane; structural transitions between the pasta phases induced by compression and elucidation of the mechanism by which they proceed. In the present article, we give an overview of the basic physics and astrophysics of the pasta phases and review our works for readers in other fields.

In: Superconductivity and Condensed Matter ...
Editor: Collin J. Eckles

ISBN 978-1-61209-785-5

Chapter 87

THE DYNAMICS OF POLYMERS IN SOLUTION WITH HYDRODYNAMIC MEMORY

V. Lisy, J. Tothova, B. Brutovsky
Institute of Physics, P.J. Safarik University, Jesenna 5, 041 54 Kosice, Slovakia
A. V. Zatovsky
Department of Theoretical Physics, I.I. Mechnikov Odessa National University, 2, Dvoryanskaya Str., 65026 Odessa, Ukraine

RESEARCH SUMMARY

The dynamics of individual polymers in solution is fundamental for understanding the properties of polymeric systems. Consequently, considerable work has been devoted towards understanding polymer dynamics. In spite of the long-standing investigations, a number of problems remains between the theory and experiments, such as the dynamic light or neutron scattering. For example, the value of the first cumulant of the dynamic structure factor (DSF) is lower than the theoretical prediction for flexible polymers. The origin of these discrepancies is a matter of continuous discussion. The development of the theory of polymer dynamics is thus still of interest.

This work represents an attempt of such a development. The main idea of the proposed generalization comes from the theory of the Brownian motion, which lies in the basis of the bead-spring models of polymer dynamics. In the classical (Einstein) description the resistance force on the bead moving in a liquid is the Stokes force, which is valid only for the steady motion. In a more general nonstationary case the hydrodynamic memory, which is a consequence of fluid inertia, should be taken into account. In the Brownian motion it displays in the famous "long-time tails" in the velocity autocorrelation function. The time dependence of the mean square displacement (MSD) of the particle changes from the "ballistic" regime at short times to the Einstein diffusion at long times. We have found similar effects in the dynamics of polymers. We give the corresponding generalization of the Rouse-Zimm (RZ) theory. It is shown that the time correlation functions describing the polymer motion essentially differ from those in the RZ models. The MSD of the polymer coil is at short times proportional to t^2 (instead of t). At long times it contains additional (to the Einstein term) contributions, the leading of which is ~ $t^{1/2}$. The relaxation of the internal normal modes of the polymer differs from the traditional exponential decay. This is displayed in the tails of their correlation functions, the longest-lived being ~ $t^{-3/2}$ in the Rouse limit and $t^{-5/2}$ in the Zimm case when the hydrodynamic interaction is taken into account. It is discussed that the found peculiarities, in particular a slower diffusion of the coil, should be observable in dynamic scattering experiments. The DSF and the first cumulant of the polymer coil are calculated. Finally, we extend the theory to the situation when the dynamics of the studied polymer is influenced by the presence of other polymers in dilute solution.

In: Superconductivity and Condensed Matter ...
Editor: Collin J. Eckles
ISBN 978-1-61209-785-5

Chapter 88

FORMATION OF MOLECULES AND ENTANGLED ATOMIC PAIRS FROM ATOMIC BEC DUE TO FESHBACH RESONANCE

Vladimir A. Yurovsky
School of Chemistry, Tel Aviv University, Tel Aviv 69978, Israel

RESEARCH SUMMARY

Bose-Einstein condensate (BEC) is considered under conditions of Feshbach res-onance in two-atom collisions due to a coupling of atomic pair and resonant molec-ular states. The association of condensate atoms can form a molecular BEC, and the molecules can dissociate to pairs of entangled atoms in two-mode squeezed states. Both entanglement and squeezing can be applied to quantum measurements and infor-mation processing.

The processes in the atom-molecule quantum gas are analyzed using two theo-retical approaches. The mean-field one takes into account deactivating collisions of resonant molecules with other atoms and molecules, neglecting quantum fluctuations. This method allows analysis of inhomogeneous systems, such as expanding BEC. The non-mean-field approach — the parametric approximation — takes into account both deactivation and quantum fluctuations. This method allows determination of optimal conditions for formation of molecular BEC and describes Bose-enhanced dissocia-tion of molecular BEC, as well as entanglement and squeezing of the non-condensate atoms.

In: Superconductivity and Condensed Matter …
Editor: Collin J. Eckles
ISBN 978-1-61209-785-5

Chapter 89

AMPHIPHILIC FLUIDS: MESOSCOPIC MODELLING AND COMPUTER SIMULATIONS

Nelido Gonzalez-Segredo
FOM Institute for Atomic and Molecular Physics (AMOLF) Kruislaan 407,
1009 DB Amsterdam, The Netherlands

RESEARCH SUMMARY

After introducing amphiphilic fluids and their technological and fundamental rel-evance, I focus attention on reviewing recent amphiphilic fluid research carried out using five different mesoscopic methods, namely, coarse-grained molecular dynamics, lattice-gas automata, lattice-Boltzmann, dissipative particle dynamics and stochastic rotation dynamics. Their *leit motif* is to pick out the atomistic detail most relevant for self-assembly and the emergence of hydrodynamics, and to map out the model's pa-rameter space. I give an account of each one's strengths and weaknesses, and suggest ways forward by pointing out phenomenology hitherto untackled by the computational modeller.

Science may be described as the art of systematic over-simplification—the art of discerning what we may with advantage omit.

—Karl Popper

In: Superconductivity and Condensed Matter ...
Editor: Collin J. Eckles
ISBN 978-1-61209-785-5

Chapter 90

DISCRETE BREATHERS IN PROTEIN SECONDARY STRUCTURE

A.E. Sitnitsky*
Institute of Biochemistry and Biophysics,
P.O.B. 30, Kazan 420111, Russia

RESEARCH SUMMARY

The role of the rigidity of a peptide chain in its equilibrium dynamics is investi-gated within a realistic model with stringent microscopically derived coupling interac-tion potential and effective on-site potential. The coupling interaction characterizing the chain rigidity and the effective on-site potentials are calculated for three main types of protein secondary structure. The coupling interaction is found to be surprisingly weak for all of them but different in character: repulsive for α-helix and anti-parallel β-sheet structures and attractive for parallel β-sheet structure. The effective on-site potential is found to be a hard one for α-helix and anti-parallel β-sheet and a soft one for parallel β-sheet. In all three types of protein secondary structures a stable zig-zag shape discrete breather (DB) associated with the oscillations of torsional (dihedral) angles can exist due to weakness of the coupling interaction. However, since the ab-sorption of far infrared radiation (IR) by proteins is known to require the existence of rather long chains of hydrogen bonds that takes place only in α-helicies, then one can conclude that the excitation of a DB in such a way is possible in α-helix and seems to be hardly possible in β-sheet structures. The interpretation of the recent experiments of Xie et al. on far IR laser pulse spectroscopy of proteins is suggested. The frequency of a DB in an the α-helix is obtained in the region of 115 cm^{-1} in accordance with these experiments.

* E-mail address: sitnitsky@mail.knc.ru

In: Superconductivity and Condensed Matter …
Editor: Collin J. Eckles
ISBN 978-1-61209-785-5

Chapter 91

ORGANIC-INORGANIC LAYERED HYBRID MATERIALS

B. Ingham[1], S.V. Chong[2] and J.L. Tallon[1,2]

[1] The MacDiarmid Institute for Advanced Materials and Nanotechnology, Victoria University of Wellington, Wellington, New Zealand

[2] Industrial Research Limited, Lower Hutt, New Zealand

RESEARCH SUMMARY

Organic-inorganic hybrid materials based on simple two-dimensional layers of tungsten oxide separated by organic molecules are merely one example of the huge variety of systems classed as "organic-inorganic hybrids". The materials in question have the potential to combine the high-electron mobility of the metal oxide layers with the flexibility of the organic layers. More importantly they identify the primary elements that might enable the formation of more complex nanostructures through self-assembling processes and the templating effect of the organic component. One such example that will be discussed here is the possibility of including magnetic ions within the inorganic layers, which is observed to have profound effects upon the magnetic behaviour of the resulting hybrids.

Tungsten oxide-organic amine hybrid materials have been synthesised via a range of methods and characterised using conventional materials characterisation techniques, such as X-ray diffraction, infrared and Raman spectroscopy, UV-visible spectroscopy and electron microscopy. The work is also complemented by ab initio calculations of the physical and electronic structure of the tungsten oxide hybrids. For compounds incorporating transition metal elements, the magnetisation was also measured. The effect of the low-dimensionality of the inorganic component is clearly seen in materials that include certain transition metals, in the form of magnetic transitions that indicate a crossover from 2-dimensional to 3dimensional behaviour.

In: Superconductivity and Condensed Matter ...
Editor: Collin J. Eckles
ISBN 978-1-61209-785-5

Chapter 92

SHAPING PROTEIN FOLDING DYNAMICS WITH NATIVE STATE'S GEOMETRY

Patrícia F N Faísca
Centro de Física Téorica e Computacional da Universidade de LisboaAv.
Prof. Gama Pinto 2, 1649-003 Lisboa Codex, Portugal

RESEARCH SUMMARY

Understanding the process according to which a linear polypeptide chain acquires its highly structured, biologically functional native fold is one of the most fundamental problems in science. Considerable insight into the physics of the folding process was gained from the study of simplified protein models and the use of computer simula-tions. The energy landscape theory of protein folding, which is largely the result of this approach, provided a theoretical framework which led to the emergence of new concepts (e.g., the transition state ensemble, the folding funnel, etc) and stimulated an invaluable synergy between theoretical and experimental research in the folding problem. An important development resulting from such an integrated approach is the key idea that the native topology has a dominant role in determining the kinetics and mechanisms of small, single domain, two-state folders. Such a surprising simplic-ity suggests in turn that the underlying physics of the folding process can indeed be described by relatively simple protein models.

In: Superconductivity and Condensed Matter … ISBN 978-1-61209-785-5
Editor: Collin J. Eckles

Chapter 93

DYNAMICS OF SEMIRIGID ROD POLYMERS FROM EXPERIMENTAL STUDIES

***George D.J. Phillies*[1] *and Kiril A. Streletzky*[2]**
[1]Department of Physics, Worcester Polytechnic Institute, Worcester, MA 01609-2280, US
[2]Department of Physics, Cleveland State University,Cleveland, OH 44115, US

RESEARCH SUMMARY

Experimental measurements of static and dynamic properties of non-ionic poly-mer solutions provide good testing grounds for existing and new models of polymer solution dynamics. We extensively studied aqueous solutions of the high molecular weight, rodlike, semiflexible polymer hydroxypropylcellulose (HPC). Here we present our systematic analysis of measurements of: a) low-shear viscosity η, b) quasi-elastic light scattering spectra of HPC solutions, including mode structure analysis at a range of temperatures and concentrations, c) quasi-elastic light scattering spectra of optical probe particles diffusing through HPC solutions, including careful characterization of modes of diffusive relaxation for tracer particles of different sizes, and d) static light scattering.

We found a variety of novel phenomena. (i) The functional dependence of η on concentration has a transition at c^+, with disparate small-and large-concentration dependences being seen. For $c < c^+$, η depends on c via a stretched exponential $\exp(-ac^\nu)$in c ("solution-like" behavior). For $c > c^+$, η depends on c via a power law $\eta \sim c^x$ with $x \approx 4$ ("meltlike" behavior). The viscometric crossover at c^+ is echoed by the probe diffusion data, confirming the physical reality of the solution-to-meltlike transition. (ii) Optical probe spectra and polymer spectra are both strongly non-exponential and can be decomposed into two or (at larger polymer concentra-tions) three spectral modes. Mode structure analysis reveals that probe relaxations and polymer relaxations in the same solution sample different aspects of polymer dy-namics. (iii) Except for the largest (0.76μm) probes, diffusion coefficients of tracer particles in polymer solutions are not determined by the solution macroscopic viscosity. (iv) At concentrations above c^*, light scattering spectra of HPC in solution have an ultraslow relaxational mode. This mode exhibits properties expected for long-lived dynamic structures but not the properties expected for local equilibrium regions of elevated polymer concentration. Properties of the slow mode are consistent with pre-dictions from some models of glass formation. Finally, (v) polymer solutions have a characteristic dynamic length, which is approximately the hydrodynamic radius of the polymer.

In: Superconductivity and Condensed Matter ...
Editor: Collin J. Eckles
ISBN 978-1-61209-785-5

Chapter 94

SOFT POLYMERIC COMPOSITES FOR ELECTRO-MECHANICAL APPLICATIONS: PREDICTING AND DESIGNING THEIR PROPERTIES BY NUMERICAL SIMULATIONS

Enis Tuncer* and Michael Wegener†
Applied Condensed-Matter Physics, Department of Physics,
University of Potsdam, D-14469 Potsdam, Germany

RESEARCH SUMMARY

Polymeric materials play an important role in daily life due to possibilities of tailor-ing special materials based on their unique properties, *e.g.*, processability, light weight, electrically/thermally insulating, chemical resistivity, mechanical softness, shock ab-sorbing *etc.* In addition to massive, large-scale, bulk applications as packaging materi-als, cables, pipes *etc.*, technological applications based on special polymers have been emerged, for example in the case of employing active thin polymeric layers to convert mechanical energy to electricity or the opposite to convert electrical energy to me-chanical stress. These electro-mechanical (also known as piezoelectric) polymers are introduced in a broad variety of electro-acoustic and electro-mechanical applications. Some examples are ultrasonic transducers in medician, defect characterization sen-sors, microphones, car air-bag sensors, shock-wave transducers, surveillance and traf-fic sensors, touch pads and panels, *etc.* The prevalent polymeric materials for electro-mechanical tranducer applications are polyvinylidene fluoride and its copolymers, and cellular polypropylene. The origin of the electro-mechanical activity in these poly-mers are quite different. The piezoelectricity in the former materials are caused by the change in their electrical dipole density, created by the

* Corresponding author. E-mail address: enis.tuncer@physics.org. Current address: Applied Superconductivity Group, Fusion Energy Division, Oak Ridge National Laboratory, Oak Ridge TN 37831-6122 USA
† E-mail address:michael.wegener@gmx.de. Current address: Fraunhofer-Institut fr Angewandte Polymer-forschung, Wissenschaftspark Golm, Geiselbergstr. 69, 14476 Potsdam Germany

oriented polar molecules, when a mechanical stress is applied; these materials belong to ferroelectrics. Un-like polyvinylidene fluoride based materials, which have been well-investigated in the last forty years, cellular polypropylene has recently shown a ferroelectric-like behav-ior. The polarizable units in cellular polypropylene are macroscopic charged voids, which generate the so-called *'ferroelectret'* behavior. One of the advantages of cellular polypropylene over others is that the internal structure of cellular films can be modified to increase the electro-mechanical activity, up-to ten times higher than polyvinylidene fluoride based materials. To better understand the physics behind and to be able to de-sign special cellular materials for transducer applications computer simulations are vi-tal. In the computer model, one should not only include the properties of constituents, *e.g.* polypropylene and the void, but also the arrangement of the constituents in the materials (the actual topology). Once these parameters are known, it is trivial to sim-ulate (predict) the behavior of the material system under different physical conditions, it would otherwise be expensive and time consuming in trial-and-error experimental techniques. In this chapter, we will first present briefly the preparation process of the cellular films, later will give the essential steps to model, to predict and to design soft-materials for electro-mechanical applications. We will first concentrate on, separately, electrical and mechanical aspects of material modeling. Later, the coupling of these stress fields and their importance in understanding the properties of materials used in the electro-mechanical applications will be presented.

In: Superconductivity and Condensed Matter ...
Editor: Collin J. Eckles
ISBN 978-1-61209-785-5

Chapter 95

ELASTICITIES AND STABILITIES: LIPID MEMBRANES VS CELL MEMBRANES

***Z.C. Tu*[1], *R. An*[2] *and Z.C. Ou-Yang*[1]**

[1]Institute of Theoretical Physics, the Chinese Academy of Sciences
Beijing 100080, China

[2]Department of MathematicsHong Kong University of Science and Technology
Clear Water Bay, Kowloon, Hong Kong

RESEARCH SUMMARY

A cell membrane can be simplified as a composite film consisting of lipid bilayer, membrane skeleton beneath the lipid bilayer, and proteins embedded in the lipid bi-layer and linked with the membrane skeleton if one only concerns its mechanical prop-erties. In this Chapter, above all, the authors give a brief introduction to some impor-tant results on mechanical properties of lipid bilayers following Helfrich's seminal work on spontaneous curvature energy of lipid bilayers. Next, the entropy of a chain-like polymer confined in a curved surface and the free energy of membrane skeleton are obtained by scaling analysis. It is found that the free energy of cell membranes has the form of the in-plane strain energy plus Helfrich's curvature energy. The equations to describe equilibrium shapes and in-plane strains of cell membranes by osmotic pres-sures are obtained by taking the first order variation of the total free energy containing the elastic free energy, the surface tension energy and the term induced by osmotic pressure. The stability of spherical cell membranes is discussed and the critical pres-sure is found to be much larger than that of spherical lipid bilayers without membrane skeleton. Lastly, the authors try to extend the present static mechanical model of cell membranes to the cell structure dynamics by proposing a group of coupling equations involving the tensegrity architecture of cytoskeleton, the fluid dynamics of cytoplasm and elasticities of cell membranes.

In: Superconductivity and Condensed Matter ... ISBN 978-1-61209-785-5
Editor: Collin J. Eckles

Chapter 96

ON THE QUANTUM NATURE OF VORTEX

Rongchao Ma
Northwest Institute for Nonferrous Metal Research, Xi'an, Shaanxi 710016, P. R. China
Yuefei Ma
School of Metallurgical Science and Engineering, Central South University, Changsha, Hunan 410083, P. R. China

RESEARCH SUMMARY

Vortices in type-II superconductors are usually considered as macroscopic entities and treated with classical mechanics because they have large size. However, a number of experimental results show that vortices can undergo quantum effects, which imply that, in some sense, the nature of vortices should be described by quantum mechanics. In this paper, the quantum effects of vortex in type-II superconductors is studied philosophically by reviewing the ideas in elementary quantum mechanics, in particular, the origin of the quantum nature of vortex is discussed and is concluded to be of that vortex has small mass per unit length. Other vortex matter phenomenology in type II superconductors, such as quantum creep, quantum depinning and quantum melting are also reviewed. It is believed that quantum effects will manifest themselves at sufficiently low temperature because in high temperature (large quantum numbers) the energy differences become arbitrarily small and the energies form almost a continuum. The quantum fluctuations of vortices drive various types of phase transitions and define much the superconducting properties of real materials at low temperature. The extension of the quantum description to the fields beyond the micro world, such as flux pinning classification and economics, is also given.

In: Superconductivity and Condensed Matter …
Editor: Collin J. Eckles

ISBN 978-1-61209-785-5

Chapter 97

CRITICAL CURRENTS AND VORTEX DYNAMICS IN PERCOLATIVE SUPERCONDUCTORS CONTAINING FRACTAL CLUSTERS OF A NORMAL PHASE

Yuriy I. Kuzmin*

Ioffe Physical Technical Institute of the Russian Academy of Sciences,
Polytechnicheskaya 26 St., Saint Petersburg 194021, Russia

RESEARCH SUMMARY

The effect of fractal clusters on magnetic and transport properties of percolative superconductors is considered. The superconductor contains percolative superconducting cluster, carrying a transport current, and clusters of a normal phase, which act as pinning centers. A prototype of such a structure is a high-temperature superconducting (HTS) wire. The superconducting core of first generation HTS wires is the conglomeration of superconducting micro-crystallites containing normal-phase clusters inside. There are the clusters of columnar defects in superconducting layer, which is the current-carrying element of second generation HTS wire. It is found that normal-phase clusters can have essential fractal features that affect the vortex dynamics. The fractal dimension of the boundary of normal-phase clusters in YBCO films is estimated and the cluster statistics is analyzed. Depinning and transport of vortices in fractal superconducting structures are investigated. Transition of the superconductor into a resistive state corresponds to the percolation transition from a pinned vortex state to a resistive state when the vortices are free to move. It is revealed that a mixed state of the vortex glass type is realized in the superconducting system involved. The fractal distribution of critical currents is derived and its peculiarities are studied. It is found that there is the range of fractal dimension where this distribution has anomalous statistical properties, specifically, its dispersion becomes infinite. The current-voltage characteristics of superconductors containing fractal clusters are obtained. Dependencies of the free vortex density on the fractal dimension as well as the resistance on the transport

* E-mail address: yurk@mail.ioffe.ru, iourk@yandex.ru, Tel.: +7 812 2479902; Fax: +7 812 2471017

current are studied. It is found that the fractality of the cluster boundary intensifies pinning and thereby raises the critical current. This feature enables the current-carrying capability of a superconductor to be enhanced without changing of its chemical composition.

In: Superconductivity and Condensed Matter ...
Editor: Collin J. Eckles
ISBN 978-1-61209-785-5

Chapter 98

APPLICATIONS OF CONFINED QUANTUM FIELDTHEORY TO CONDENSED MATTER SYSTEMS: SUPERCONDUCTING PHASE TRANSITIONS IN FILMS, WIRES AND GRAINS

***A. P. C. Malbouisson*[1*], *J. M. C. Malbouisson*[2†] *and A. E. Santana*[3‡]**

[1]Centro Brasileiro de Pesquisas F´ısicas (CBPF/MCT), Rua Dr. Xavier Sigaud, 150, Rio Janeiro, 222290-180, RJ, Brazil

[2]Instituto de Física, Universidade Federal da Bahia,Salvador, 40210-340, BA, Brazil

[3]Instituto de Física, Universidade de Brasília,Brasília, 70910-900, Brazil

RESEARCH SUMMARY

We consider the $\lambda\varphi^4$ theory in the D-dimensional Euclidean space with $d(d \leq D)$ compactified dimensions. Parameterizing the mass term in the form $m^2(T) = \alpha(T - T_0)$, we retrieve Ginzburg-Landau (GL) models with the order parameter confined in d spatial dimensions. We use methods of Quantum Field Theory to discuss second-order phase transitions in such framework. Specifically, we consider the 3-dimensional GL model compactified in $d = 1, 2$ and 3 space coordinates describing a film, a wire and a grain, respectively. We determine the dependence of the transition temperature with the relevant size of the system for these cases. In particular, we address the question of the existence of a minimal size of the system for sustaining the superconducting phase. We also consider the critical behavior of type-II superconducting films in a magnetic field. The results are in qualitative agreement with the experimental measures for the case of superconducting transitions.

* E-mail address: adolfo@cbpf.br
† E-mail address: jmalboui@ufba.br
‡ E-mail address: asantana@fis.unb.br

In: Superconductivity and Condensed Matter ...
Editor: Collin J. Eckles

ISBN 978-1-61209-785-5

Chapter 99

COMBINED EFFECTS OF DISORDER AND MAGNETIC FIELD IN SUPERCONDUCTORS

P.D. Sacramento and J. Lages

Centro de Fí sica das Interacc¸oes Fundamentais, Ĩnstituto Superior Técnico,
Av. Rovisco Pais, 1049-001 Lisboa, Portugal
Laboratoire de Physique Théorique, UMR 5152 du CNRS, Université Paul Sabatier, 31062 Toulouse Cedex 4, France

RESEARCH SUMMARY

We consider the effects of disorder due to impurities acting as scatterers for the electrons and as pinning centers for vortices, induced by an external magnetic field, in a strongly type-II superconductor. We consider the cases of s-wave and d-wave pairing symmetries. The effect of disorder reflects in the density of states, particularly at low energies. Considering first that no impurities are present but that the vortices are pinned randomly, the density of states as a function of magnetic field increases at low energies, closing the gap in the s-wave case and becoming finite in the d-wave case. Including the effect of the scattering from impurities, placed randomly using the binary alloy model, the density of states and the local density of states (LDOS) have to be solved self-consistently. The results for the LDOS agree qualitatively with experimental results if most vortices are pinned to the impurities. Also, we consider the effect of impurities on the vortex charge.

In: Superconductivity and Condensed Matter …
Editor: Collin J. Eckles
ISBN 978-1-61209-785-5

Chapter 100

ROOM TEMPERATURE RADIOFREQUENCY ELECTRODYNAMIC PROPERTIES OF BULK LANTHANUM MANGANITES

A. Rinkevich [*] ***and A. Nosov***
Institute of Metal Physics, Ural Division of the Russian Academy of Sciences
18 Sofia Kovalevskaya St., GSP-170, Ekaterinburg 620041, Russia

RESEARCH SUMMARY

Due to the effect of "colossal" magnetoresistance lanthanum manganites have become one of the most fashionable objects of investigations in solid state physics. Most publications devoted to investigations of physical properties of doped lanthanum manganites report about the results obtained by static (under the *dc* excitation current or at frequencies below 1 kHz) methods. Investigations of properties of doped manganites in the radio-frequency range (from 10 kHz to 100 MHz) received much less attention. Meanwhile in the case of manganites the range of radio frequencies can be of particular interest since the magnitude of effects observed can be much higher than under *dc* current excitation. The value of skin depth can be tailored by variation of frequency and can be comparable with the size of a sample. In such case different mechanisms of interaction between an electromagnetic wave and spin subsystem of a sample under investigation can be analyzed in detail. When the thickness of a sample exceeds the value of skin depth variations of the penetration coefficient in the *dc* magnetic field are mainly determined by variation of impedance of a sample. Variations of impedance are caused by resistivity and dynamic magnetic permeability. Therefore, important information about dynamic magnetic properties of doped manganites can be obtained from investigations of penetration of electromagnetic field.

---

[*] E-mail address: rin@imp.uran.ru

In: Superconductivity and Condensed Matter …
Editor: Collin J. Eckles
ISBN 978-1-61209-785-5

Chapter 101

CURRENT INSTABILITY INVESTIGATIONS OF CONDUCTION-COOLED HIGH TEMPERATURE SUPERCONDUCTING COMPOSITES

V. R. Romanovskii*[*,a] and K. Watanabe[b]
[a]Russian Research Center 'Kurchatov Institute', Moscow 123182, Russia
[b]High Field Laboratory for Superconducting Materials,
Institute for Materials Research, Tohoku University, Sendai 980-8577, Japan

RESEARCH SUMMARY

A current instability problem of a conduction-cooled high-T_c superconducting composite is discussed in detail. Transient and static approximations are used to define the boundary between stable and unstable currents charged in a composite. The corresponding quench electric field and current are derived under the aspect of the possible finite temperature rise of a composite that is unavoidable before current instability. It is shown that the allowable temperature variation of a composite is not trivial and depends on the properties of the superconductor and matrix. In the limiting case, a stable temperature rise may occur up to the critical temperature when the complete thermal stability condition takes place. The latter is formulated for a superconducting composite with the power law of the voltage-current characteristic.

The performed analysis allows to reveal that stable and unstable values of the electric field and current might be both sub-critical and over-critical, i.e. lower and higher of the corresponding values following from the fixed electric field criterion.

The criteria describing the existence of such operating states are formulated. Besides, it is stated that the allowable stable temperature change of a composite leads to the thermal degradation of its quench parameters. As a result, the quench currents do not increase proportionally with relevant increase of the superconducting critical current. It is shown that operating regimes may have many-valued stable branches appearing in accordance with

[*] E-mail address: vromanovskii@netscape.net

nonlinear temperature dependence of the critical current of a superconductor. The parameters which influence on these current stability mechanisms are defined.

In: Superconductivity and Condensed Matter ...
Editor: Collin J. Eckles
ISBN 978-1-61209-785-5

Chapter 102

RADIATION SHIELDING SCHEMES AND ADVANCED FABRICATION TECHNIQUES FOR SUPERCONDUCTING MAGNETS

***Laila A. El-Guebaly*[1,*] *and Lester M. Waganer*[2,♣]**
[1]University of Wisconsin, Fusion Technology Institute, Madison, WI 53706-1609, US
[2]Consultant for the Boeing Company, US

RESEARCH SUMMARY

The tokamak has been perceived as the most promising magnetic confinement concept to achieve fusion energy. For economic reasons, superconducting rather than normal magnets will be used in commercial magnetic fusion plants to significantly reduce the circulating power fraction. A number of superconducting magnet designs have been developed over the past 50 years with both low and high temperature superconductors. For a practical fusion facility, these magnets must be well shielded from energetic fusion neutrons to achieve the predicted performance and service lifetime. With innovative manufacturing techniques, these magnets can be fabricated faster, cheaper, and more practically than conventional processes currently indicate.

* E-mail address: elguebaly@engr.wisc.edu.
♣ E-mail address: lesw@centurytel.net.

In: Superconductivity and Condensed Matter ... ISBN 978-1-61209-785-5
Editor: Collin J. Eckles

Chapter 103

DIPOLON THEORY OF KINK STRUCTURE OF QUASI-PARTICLE ENERGY DISPERSION OBSERVED IN PHOTOEMISSION SPECTRA OF HIGH TEMPERATURE SUPERCONDUCTERS $BI_2SR_2CACU_2O_{8+\Delta}$

R. R. Sharma*
Department of Physics, University of Illinois at Chicago, Chicago, IL 60607-7059, US

RESEARCH SUMMARY

Formerly unexplained experimentally deduced kink in the quasiparticle energy dispersion observed in photoemission spectra of high-temperature superconducters $Bi_2Sr_2CaCu_2O_{8+\delta}$ has been explained here naturally via dipolon mediated electron-electron pairing mechanism of high-temperature superconductors. We have made use of the four-momenta space diagrams in the dressed particle picture to write the self-energy $\Sigma(p)$by taking the sum of the exchange diagrams involving dipolon propagator, electron Green's function and screened electron-electron Coulomb repulsion taking into account explicitly the dressed dipolons as mediators of superconductivity and nonrigid electron energy bands considering retardation and damping effects and electron-hole asymmetry to obtain single-quasiparticle energy dispersion in high-$T_cBi_2Sr_2CaCu_2O_{8+\delta}$ superconductors. The theory contains Mott renormalization, all important and necessary electron correlations and inherently the interaction on the superconducting layer due to all the planes of atoms and chains of atoms in the system. This constitutes an extension of the strong-coupling dipolon theory which explains also the regular and the pseudo-energy gaps, T_c, T_c* and the peak-dip-hump structure of the line shape of the photoemission spectra of highT_c superconductors. Our calculations of the single-quasiparticle

* E-mail address: sharma@uic.edu; University of Illinois at Chicago, 845 W. Taylor Street, Rm. 2236 (MC 273), Chicago, IL 60607-7059; Tel.#: (312)996-3400; Fax#: (312)996-9016

energy dispersion for$Bi_2Sr_2CaCu_2O_{8+\delta}$ obtain a low energy kink at the binding energy near 60 meV which has already been identified in the experiments and also predict two additional high energy kinks at binding energies close to 110 and 170 meV, yet to be identified experimentally, in confirmation with the previous predictions by dipolon theory. The calculations not only explain the observed velocities of the quasiparticles below and above the low energy kink but also predict the velocities below and above the other predicted high energy kinks. The origin of the kinks and the effect of doping, temperature, interaction with phonons and coupling with magnetic modes have been discussed. The Migdal vertex correction does not change our results drastically. Direct relation between the photoemission spectra and the kink structure has been revealed. The knowledge gained from our calculations and explanation of the peak-dip-hump phenomenon and the kink structure has been explicitly written down in the form of five principles for understanding the photoemission results in high-T_c superconductors. The predicted results from our calculations for the high energy kinks, the quasiparticle velocities and the quasiparticle density of states are proposed to be observed by the experiments.

In: Superconductivity and Condensed Matter ... ISBN 978-1-61209-785-5
Editor: Collin J. Eckles

Chapter 104

ANALYTICAL MODEL OF THERMAL STRESSES AND ANALYTICAL FRACTURE MECHANICS IN YBACUO SUPERCONDUCTOR

Ladislav Ceniga
Institute of Materials Research, Slovak Academy of Sciences,
Watsonova 47, 040 01 Kosice, Slovakia

RESEARCH SUMMARY

The chapter deals with analytical modelling of elastic thermal stresses in an isotropic-anisotropic multi-particle-matrix system consisted of periodically distributed isotropic spherical particles in an uniaxial-anisotropic in£nite matrix to be imaginarily divided into identical cells containing a central spherical particle. The multi-particle-matrix system represents a model system applicable to real two-component materials of a precipitation-matrix type. Consequently, the cell as a part of the in£nite matrix related to one spherical particle is characterized by particle volume fraction, inter-particle distance and particle radius, representing microstructural parameters of the real two-component materials. Investigated within the cell, the thermal stresses as functions of the microstructural parameters originate during a cooling process as a consequence of the difference in thermal expansion coef£cients of the matrix and the particle. The analytical modelling as the £rst topic of the chapter results from the Cauchy's, Saint-Venant's, equilibrium equations, and the Hooke's law, representing fundamental equations of Continuum Mechanics. In addition to analytical modelling of the thermal stresses, the second topic of the chapter is represented by analytical fracture mechanics to include analytical modelling of the crack initiation and the crack propagation related to any and ideal-brittle multi-particle-matrix system, respectively. The analytical fracture mechanics results from 'curve' elastic energy density as an in-tegral of thermal-stress induced elastic energy density along a curve in the cell. Con-sequently, the 'curve' elastic energy density results in the determination of critical particle radius as a consequence of the crack initiation as well as in the determination of a function describing the crack shape and the crack dimension in a plane perpendic-ular to the direction of the crack propagation. The analytical results presented in the chapter are applied to the YBaCuO superconductor as the Y2BaCuO5-YBa2Cu3O7 multi-particle-matrix system to consist of the Y2BaCuO5 isotropic particle and the YBa2Cu3O7 uniaxial-anisotropic matrix, both representing ceramic components con-sidered to be ideal-brittle regarding the analytical modelling.

In: Superconductivity and Condensed Matter ... ISBN 978-1-61209-785-5
Editor: Collin J. Eckles

Chapter 105

SUPERCONDUCTING FLUCTUATIONS ABOVE T_C IN OVERDOPED $Bi_2Sr_2CaCu_2O_{8+\delta}$

***Z. A. Xu*[1], *J. Q. Shen*[1], *Z. W. Zhu*[1], *S. Ooi*[2], *T. Mochiku*[2] *and K. Hirata*[2]**
[1]Department of Physics, Zhejiang University, Hangzhou 310027, China
[2]National Institute for Materials Science, 1-2-1 Sengen, Tsukuba, Ibaraki 305-0047, Japan

RESEARCH SUMMARY

The resistivity, Hall effect and Nernst effect of overdoped $Bi_2Sr_2CaCu_2O_{8+\delta}$ single crystals were studied to investigate superconducting fluctuations above T_c. The superconducting transition temperature (T_{c0}) for the three measured samples with various oxygen content is 86.5, 83 and 77 K, and their doping level (hole concentration, p) is estimated to be 0.180, 0.193 and 0.204 respectively. The pseudogap opening temperature (T^*) determined from the temperature dependence of resistivity decreases slightly from 126 K to 100 K with increasing doping level p. The temperature scale (T_1) determined from the cotangent of Hall angle ($\cot\theta$) is consistent with the onset temperature (T_v) of vortex-like excitation detected by Nernst effect, and both decrease with increasing p. Moreover, none of the three temperature scales (T^*, T_1 and T_v) drops to zero as p increases across the proposed quantum critical point (QCP) of p ~0.19. This result indicates that both pseudogap opening and superconducting fluctuations exist above T_c in the overdoped region of electronic phase diagram of cuprates and there is no evidence for QCP at p ~ 0.19. We suggest that the temperature range between T_c and T_v (or T_1) is characterized by finite pairing amplitude with strong phase fluctuations. As T is above T_v, the phase correlation disappears completely and only the pairing amplitude fluctuations remain. The strong superconducting phase fluctuation in the pseudogap regime implies that the pseuodgap is a precursor to superconductivity.

In: Superconductivity and Condensed Matter ...
Editor: Collin J. Eckles
ISBN 978-1-61209-785-5

Chapter 106

VORTEX ENTRY CONDITIONS IN TYPE-II SUPERCONDUCTORS: GIBBS-LONDON APPROACH VS GINZBURG-LANDAU THEORY

I. L. Maksimov*
Nizhny Novgorod University, 603950, Nizhny Novgorod, Russian Federation

RESEARCH SUMMARY

The problem of the vortex entry into type-II superconductors (bulk samples, thin films) is analysed theoretically. Two approaches for the problem are discussed:

i) the Gibbs-London theory, based on the concept of geometrical barrier, controlling vortex entry/exit into/out from superconductor, and
ii) the quantum-mechanical Ginzburg-Landau theory, employing the concept of the wave function of superconducting condensate.

The quantitative conditions for the entry of vortices into type-II superconductors being in the Meissner and/or mixed state, will be described in the framework of two different approaches. Specifically, a detailed theory of geometrical barrier will be constructed, on the basis of which the shape of the vortex fragment will be described, and the vortex-entry criterion will be obtained. The analytical and numerical study of the Ginzburg-Landau equations reveal the existence of a modulation instability of superconducting condensate. The instability threshold is reached when superconducting order parameter turns to zero at the sample surface and the supervelocity (or gauge-invariant vector potential) of the condensate reaches some a threshold value at the superconductor edge or surface. For superconductors with high value of Ginzburg-Landau parameter this criterion practically coincides with criterion based on the depairing current density. Due to this instability, vortices start to nucleate at the edge or surface and then penetrate deeper into the sample.

* E-mail address: ilmaks@phys.unn.ru

In: Superconductivity and Condensed Matter ...
Editor: Collin J. Eckles

ISBN 978-1-61209-785-5

Chapter 107

PROGRESS IN FABRICATION TECHNIQUES FOR SMALL-CAPACITANCE JOSEPHSON JUNCTIONS

***Michio Watanabe*[1], *Kazuaki Matsuba*[3,4], *Yasunobu Nakamura*[1,2,4] *and Jaw-Shen Tsai*[1,2,4]**

[1]Frontier Research System, RIKEN, Saitama 351-0198, Japan
[2]NEC Fundamental and Environmental Research Labs., Ibaraki, 305-8501, Japan
[3]Tokyo Institute of Technology, Yokohama, 226-8502, Japan
[4]CREST-JST, Saitama 332-0012, Japan

RESEARCH SUMMARY

We have developed a fabrication process for small-capacitance tunnel junctions, which includes focused-ion-beam (FIB) etching from different directions. By apply-ing the process to a Nb/(Al–)Al_2O_3/Nb trilayer, we have fabricated single-electron transistors (SETs), and characterized the SETs at low temperatures, $T = 0.04 - 40$ K. The superconducting gap energy and the transition temperature of the SETs agree with the bulk values of Nb, which suggests high-quality Nb/(Al–)Al_2O_3/Nb junctions. The effective area of the junctions is $<0.1 \times 0.1$ μm^2, and the single-electron charging energy of the SETs is estimated to be larger than 1 K. In this chapter, we review ear-lier attempts to fabricate small-capacitance Josephson junctions with Nb electrodes, and describe our nanoscale fabrication process in detail. We also discuss the transport properties of our SETs by comparing with the theoretical predictions.

In: Superconductivity and Condensed Matter ...
Editor: Collin J. Eckles
ISBN 978-1-61209-785-5

Chapter 108

ON THE QUANTUM NATURE OF VORTEX

Rongchao Ma
Northwest Institute for Nonferrous Metal Research, P. O. Box 51, Xi'an, Shaanxi 710016, P. R. China
Yuefei Ma
School of Metallurgical Science and Engineering, Central South University, Changsha, Hunan 410083, P. R. China

RESEARCH SUMMARY

Vortices in type-II superconductors are usually considered as macroscopic entities and treated with classical mechanics because they have large size. However, a number of experimental results show that vortices can undergo quantum effects, which imply that, in some sense, the nature of vortices should be described by quantum mechanics. In this paper, the quantum effects of vortex in type-II superconductors is studied philosophically by reviewing the ideas in elementary quantum mechanics, in particular, the origin of the quantum nature of vortex is discussed and is concluded to be of that vortex has small mass per unit length. Other vortex matter phenomenology in type II superconductors, such as quantum creep, quantum depinning and quantum melting are also reviewed. It is believed that quantum effects will manifest themselves at sufficiently low temperature because in high temperature (large quantum numbers) the energy differences become arbitrarily small and the energies form almost a continuum. The quantum fluctuations of vortices drive various types of phase transitions and define much the superconducting properties of real materials at low temperature. The extension of the quantum description to the fields beyond the micro world, such as flux pinning classification and economics, is also given.

Chapter 109

ELECTRIC NOISE AND LOCAL PHOTON-INDUCED NONEQUILIBRIUM STATES IN A CURRENT-CARRYING NANOSTRUCTURED SUPERCONDUCTOR

Andreas Engel
Physics Institute of the University of Zurich, Winterthurerstr, 190, 8057 Zurich, Switzerland
Alexei Semenov and Heinz-Wilhelm Hübers
DLR, Institute of Planetary Research, Rutherfordstr. 2, 12489 Berlin, Germany
Konstantin Il'in and Michael Siegel
Institute for Micro-and Nanoelectronic Systems, University of Karlsruhe, Hertzstr, 16, 76187 Karlsruhe, Germany

RESEARCH SUMMARY

Superconducting thin films and micro-and nanostructured superconducting elements have enormous potential for applications in electronics and as highly sensitive detectors. We present recent studies of the superconductivity of nanostructured Nb and NbN thin films as well as their potential use as very fast, sensitive and energy-dispersive photon-counters. Using photo-and electron-beam-lithography 5 to 20 nm thick films were patterned into single bridges and meander structures with widths of the conduction paths as narrow as 50 nm. Dependencies of the critical temperature and critical current density on variations in bridge width and thickness can be modelled rather well using existing theories applicable to dirty superconductors with very short coherence lengths.

The meander structures have already been shown to be very fast and sensitive single photon detectors in the visible and near-infrared range. Their ultimate sensitivity is hampered however, by electric noise caused at least partly by fluctuations in the superconducting state. The detection mechanism exploits the creation of a localized hot-spot by the absorption of a photon with sufficient energy. Albeit being a relatively small perturbation of the superconducting state, it can be detected as a voltage transient given the right experimental conditions. These conditions require cross-sectional areas of the conduction paths comparable with the cross-section of the hot-spot and an applied bias current only slightly less than the critical current at the operating temperature. Under these conditions fluctuations, like e.g. phase-slip centers, fluctuations in the number density of Cooper-pairs or fluctuations within the vortex system, may also cause voltage pulses very similar in magnitude and shape to those caused by absorbed photons and may be a significant contribution to the dark count rate. Measured dark count rates have been analyzed with respect to these various noise sources. Besides the noise characteristics or noise equivalent power the performance of the single-photon detectors was investigated concerning the quantum efficiencies as a function of the incident photon energy and material parameters of the superconducting meander line.